W. J. FRASER & Co. Ltd.

Chemical Engineers

Dagenham, Romford
ESSEX

Manufacturers of

NITRATORS—in Iron or Lead.

COOLING COILS—in Iron or Lead.

EVAPORATING AND CONCENTRATING APPARATUS

LEAD TANKS AND COILS

MIXERS FOR ACID OR ALKALI

STILLS—in Iron or Lead

STEAM JACKETED PANS

AIR COMPRESSORS

STORAGE TANKS—Iron or Lead

ACETONE PLANTS

ACID RECOVERY PLANTS

ACID CONCENTRATING PLANTS

INDUSTRIAL
NITROGEN COMPOUNDS
AND EXPLOSIVES

INDUSTRIAL
NITROGEN COMPOUNDS
AND
EXPLOSIVES

A Practical Treatise on the Manufacture, Properties, and
Industrial Uses of Nitric Acid, Nitrates, Nitrites, Ammonia,
Ammonium Salts, Cyanides, Cyanamide, Etc. Etc.

INCLUDING THE

MOST RECENT MODERN EXPLOSIVES

BY

GEOFFREY MARTIN

*Ph.D., D.Sc., B.Sc., F.C.S., Industrial Chemist
and Chemical Patent Expert*

AND

WILLIAM BARBOUR

M.A., B.Sc., F.I.C., F.C.S., Explosives Chemist

LONDON
CROSBY LOCKWOOD AND SON
7 STATIONERS' HALL COURT, LUDGATE HILL, E.C.
and 5 BROADWAY, WESTMINSTER, S.W.
1915

PRINTED AT
THE DARIEN PRESS
EDINBURGH

PREFACE

THE enormous and rapidly increasing importance of nitrogen compounds can scarcely be over-estimated at the present time. They form the raw materials from which are manufactured practically all the effective explosives now in use. Exceedingly large quantities of nitrogen compounds are consumed annually by the agricultural industries, in order to restore to the soil the nitrogen extracted therefrom by growing crops. Without nitrogenous manure soil becomes barren. In fact, in recent years, the agricultural industries have absorbed over 80 per cent. of the available supplies of nitrogenous materials, and the supply does not equal the demand. This ever-increasing want has led within the last few years to the creation of enormous new industries—like the cyanamide industry—in which products scarcely known to the scientist of a few years ago are now placed on the market in quantities running into hundreds of thousands of tons annually.

Within recent years methods have been worked out for converting the practically unlimited supplies of free nitrogen in the atmosphere into nitric acid, nitrates, ammonia, cyanamide, and other valuable chemicals and fertilising agents.

These achievements of the chemist form one of the romances of science, and are undoubtedly destined to bring about economic revolutions which will profoundly modify the whole surface of our planet. By means of these new industries Germany has rendered herself practically independent of external supplies of nitrates and other raw materials for making explosives.

Great Britain, however, is still almost entirely dependent upon nitrate importation from overseas, and in the interests of national safety the Government should undertake the establishment of national works for producing nitric acid and nitrates from atmospheric air by the processes which have been worked successfully abroad.

In the present volume the methods of manufacture and the utilisation of the chief industrial nitrogenous substances are given as concisely as possible.

Special weight has been laid upon new and profitable methods of production; for this reason the chief patents relating to the newer processes of manufacture have been indicated wherever possible.

For the same reason special prominence has been given to the new synthetic processes for making ammonia directly from its component elements, and its subsequent oxidation to nitric acid and ammonium nitrate by catalytic processes, which are now being worked in Germany on a very large scale.

On account of the importance of explosives at the present time, and the fact that their manufacture intimately depends upon that of nitric acid and nitrates, it has been thought advisable to include the manufacture and properties of the chief explosives now in use.

This section was contributed by Mr William Barbour, M.A., B.Sc., whose extensive experience in the manufacture of these products will make the details given of special value to practical men at the present time. The large number of patents and new processes referred to in the text are excused by the fact that although many of them have proved commercially unsuccessful in peace times, yet they often embody fruitful ideas, and under the present abnormal conditions may often be worked successfully.

The editor wishes to thank Mr J. Louis Foucar, B.Sc., late assistant manager to the Beckton Gas Works, for his kind help in revising the sections relating to ammonia and nitrates.

July 1915.

TABLE OF CONTENTS

CHAPTER I
THE CIRCULATION OF NITROGEN IN NATURE

PAGE

Literature—The Nitrogen Cycle—Elimination of Nitrogen from Plants and Animals—Nitrification —Denitrification - - - - - - - - - - - - - 3

CHAPTER II
THE NITRATE INDUSTRY

Literature—**Sodium Nitrate or Chile Saltpetre**—Occurrence—Extraction—Properties—Statistics —Uses. **Potassium Nitrate**—Manufactures—Properties—**Calcium Nitrate**—**Ammonium Nitrate—Sodium Nitrite** - - - - - - - - - 9

CHAPTER III
THE NITRIC ACID INDUSTRY

Literature—**Nitric Acid**—Manufacture from Chile Saltpetre—Manufacture from the Atmosphere— Birkeland-Eyde Furnace—Pauling Furnace—Schönherr Furnace—Other furnaces—General Plant for the Manufacture of Nitrate by Electrical Oxidation of the Air—Manufacture of Nitric acid from Ammonia (Ostwald's Process)—Properties of Nitric Acid—Statistics - 17

CHAPTER IV
THE AMMONIA AND AMMONIUM SALTS INDUSTRY

Literature—**Ammonia and Ammonium Salts**—Sources of Ammonia—Statistics—Manufacture of Ammonium Sulphate from Gas-Water or Ammoniacal Liquor—Ammonia Stills—Treatment of Waste Exit Gases from the Ammonium Sulphate Plant—Manufacture of Ammonium Sulphate from Mond Gas—The Direct Process of making Ammonium Sulphate—Kopper Ammonia Recovery Plant—Otto-Hilgenstock Ammonia Recovery Process—The Coppee Company's Semi-Direct Process - - - - - - - - 35

Manufacture of Caustic Ammonia (Liquor Ammonia)—Concentrated Ammonia Water—Manufacture of Pure Aqueous Solutions of Ammonia—Specific Gravity of Ammonia Solutions - 45

Technical Ammonium Salts—Ammonium Sulphate—Ammonium Chloride—Ammonium Carbonate — Ammonium Nitrate — Ammonium Perchlorate — Ammonium Phosphate— Ammonium Persulphate — Ammonium Thiosulphate—Ammonium Acetate—Ammonium Fluoride—Ammonium Sulphocyanide — Ammonium Chlorate — Ammonium Bromide— Ammonium Oleate—Dry Ammonia — Solid Ammonia—Anhydrous Ammonia (Liquid Ammonia) - - - - - - - - - - - 48

CHAPTER V
SYNTHETIC AMMONIA

Literature—Ammonia by Direct Union of Nitrogen and Hydrogen by Means of a Catalyst— Ammonia from Cyanamide—Ammonia from Nitrides—Serpek Process—Other Processes - 53

CHAPTER VI
THE CYANAMIDE INDUSTRY

Literature—**Calcium Cyanamide** (Nitrolime, Kalkstickstoff) — Manufacture—Uses —**Dicyandiamide—Urea—Ferrodur**—Nitrogen Products Derivable from Calcium Cyanamide - 61

CHAPTER VII

THE CYANIDE AND PRUSSIATE INDUSTRY

PAGE

Literature— Statistics — **Manufacture of Sodium and Potassium Cyanide** — From Ferro-cyanides (Prussiate)—From Ammonia, Carbon, and Alkali Metal or Alkali Salt—Sieper-mann's Process—Bielby Process—Castner Process—Manufacture of Cyanide from Sulpho-cyanides—Cyanides from "Schlempe" or Sugar Residues—Bueb's Process—Manufacture of Cyanides from Calcium Cyanamide — Manufacture of Cyanides from Atmospheric Nitrogen -   71

Ferrocyanides—**Potassium Ferrocyanide**—Properties—Old Process of Manufacture—Modern Processes of Manufacture from Coal Gas -   79

Potassium Ferricyanide—**Prussian Blue**—**Sulphocyanides or Thiocyanates**—Recovery from Coal Gas—Synthetic Sulphocyanides from Carbon Disulphide and Ammonia   81

CHAPTER VIII

THE MANUFACTURE OF NITROUS OXIDE
(Laughing Gas, Nitrogen Monoxide)

Processes of Manufacture—Properties—Analysis—Transport   84

CHAPTER IX

THE MODERN EXPLOSIVES INDUSTRY

Literature—Nature of Modern Explosives   87

Gunpowders—Old Black Powder—Bobbinite—Sprengsalpeter—Petroklastite   87

Wetterdynammon—Amide Powder   88

Nitroglycerine—Manufacture—Process of Nitration—Washing Process—Properties—Lowering the Freezing Point of   88

Dinitro-Glycerine (Glycerine Dinitrate)   91

Nitrate of Polymerised Glycerine—Dinitromonochlorhydrin—Dinitroacetin—Dinitroformin   91

Gun-Cotton—Raw Materials—The Nitrating Acids—Process of Nitration—Stabilisation—Properties   92

Dynamite, Blasting Gelatine, Gelatine Dynamite, and Gelignite—Manufacture—Properties—Method of Using   97

Picric Acid—Manufacture from Phenol—Uses—Dangers—Melinite—Lyddite—Pertite—Ecrasite—Schimose -   100

Trinitro-toluene—Process of Manufacture—Properties—Plastrotyl—Triplastit—Tolit—Trilit   101

Trinitrobenzene—**Nitronaphthalenes**—**Nitroanilines**—Flürscheim's Patents   102

Ammonium Nitrate Explosives. Safety Explosives—Nature of Safety Explosives—Method of Testing Explosives for Safety—Chief Explosives containing Ammonium Nitrate—Ammonium Nitrate Explosives exhibiting Plasticity—Ammonal   102

Chlorate and Perchlorate Explosives—Cheddites—Colliery Steelite—Silesia Powder—Permonite—Alkalsite—Yonckite   104

Tetranitromethane—Use as an Explosive   105

Fulminate of Mercury, Azides, Primers, Percussion Caps, Detonators—**Mercury Fulminate**—Preparation—Properties—**Detonators**—**Tetryl**—**Sodamide**—Preparation—Properties   105

Smokeless Powders—Chief Constituents of Smokeless Powders—**Cordite**—Manufacture—Properties—Uses—Robertson and Rintoul's Process for Recovering Acetone from Cordite Drying Houses—**Solenite**—**Ballistite**—**Sporting Powders**—Methods of Reducing Flame on Smokeless Powders—Stability of Smokeless Powders   106

Observations on the Testing of Explosives—Proximate Analysis—Stability Tests   110

Composition of Explosives in Common Use   112

Statistics Relating to Explosives   115

INDEX -   119

CHAPTER I

—

Circulation of Nitrogen in Nature

CHAPTER I

CIRCULATION OF NITROGEN IN NATURE

LITERATURE

CROSSLEY.—*Pharm. Journ.*, 1910, **30**, 329.
FOWLER.—" Bacteriological and Enzyme Chemistry." 1911.
BRAUN.—Abegg's " Handbk. d. anorg. Chem," vol. 3, pt. 3, 216.

NITROGEN is an essential component of animal and vegetable living matter, and must be supplied to them in an assimilable form.

Animals obtain their nitrogen in an "organically combined" form from vegetable matter. Plants, however, obtain their nitrogen principally in the form of nitrates or ammoniacal salts, which they absorb from the soil. To a much smaller extent they take up (at least some varieties do) nitrogen directly from the atmosphere.

When animal and vegetable matter decays, the combined nitrogen passes again in the form of ammonia and nitrates into the soil. Part, however, is set free as nitrogen gas, which thus escapes into the atmosphere.

According to Arrhenius (" Das Werden der Welten," pp. 130, 131, 1908) no less than 400,000,000 tons of nitrogen are annually withdrawn from the atmosphere either by direct assimilation by plants or by union with oxygen by means of electrical discharges which occur in the air. The combined nitrogen is washed into the soil and there taken up by plants, whence it passes into circulation in the animal world. (The amount of atmospheric oxygen annually absorbed by plants and animals from the atmosphere is about 100 times greater than the amount of nitrogen absorbed.) Now, since nitrogen does not accumulate to any great extent in the soil, these 400,000,000 tons of atmospheric nitrogen are yearly set free again as inert nitrogen gas by the decomposition of organic matter, and so are restored again to the atmosphere. Consequently there is continually going on in nature an immense and endless circulation of nitrogen.

The following scheme shows the course of the **nitrogen cycle** in nature :—

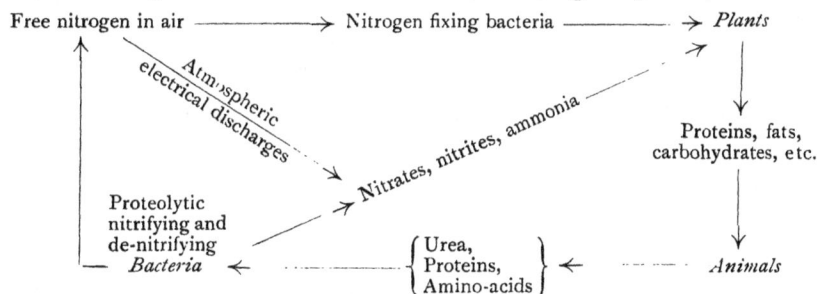

Every scrap of nitrogen in every plant and every animal, and in the soil, came originally from free atmospheric nitrogen. The fixation occurs principally by means of silent electrical discharges, which continually go on in the atmosphere. Under the influence of electricity the atmospheric nitrogen becomes active, and unites with the oxygen of the air to form nitric oxide, thus $N_2 + O_2 = 2NO$. The nitric oxide, NO, at once combines with the oxygen of the air to form nitrogen peroxide, NO_2, thus $NO + O = NO_2$. The NO_2 is then dissolved by falling rain, forming nitrous and nitric acid $(2NO_2 + H_2O = HNO_2 + HNO_3)$, and in this form it enters the soil, where the nitric acid and nitrous acid combine with bases in the soil like lime, potash, etc., to form calcium or potassium nitrate or nitrite; these substances are finally absorbed by the roots of growing plants. This is the main method whereby nitrogen is absorbed from the atmosphere, and for many years it was thought that this was the only way of absorbing nitrogen. However, it is now known, thanks to the researches of Hellriegel and Wilfarth, that certain plants, especially those belonging to the Leguminosæ, directly absorb nitrogen by means of certain bacteria. Peas and beans possess this capacity highly developed. Clover also is a plant which absorbs large amounts of nitrogen from the air during its growth. Practical examples of utilising this fact have long been known. After a crop has been grown which tends to exhaust nitrogen from the soil (*e.g.*, wheat) it is usual to grow a crop of clover, which is then ploughed into the soil. The nitrogen absorbed by the clover from the atmosphere is thus returned to the soil when the clover rots, and so a fresh crop of wheat can now be grown. This is known as "rotation of crops."

The nitrogen as nitrate in the soil is absorbed by the plant and turned into complex protein matter. The plant is eaten by animals, and the nitrogenous protein is, by the processes of peptic and tryptic digestion in the animal body, converted into end products, largely consisting of amino acids. These amino acids are again built up into the body substance through biotic energy inherent in the cells, part having been utilised as fuel in maintaining that energy.

In the case of **flesh-eating mammals** most of the nitrogen not used in maintaining the body substance is eliminated in the form of **urea**, $CO(NH_2)_2$, in the urine. From this urea much nitrogen is set free in the free state by the action of nitrous acid, HNO_2, thus :—

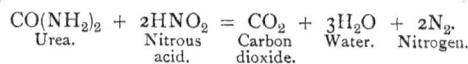

$$\underset{\text{Urea.}}{CO(NH_2)_2} + \underset{\substack{\text{Nitrous} \\ \text{acid.}}}{2HNO_2} = \underset{\substack{\text{Carbon} \\ \text{dioxide.}}}{CO_2} + \underset{\text{Water.}}{3H_2O} + \underset{\text{Nitrogen.}}{2N_2}.$$

Thus the nitrogen, in part at least, finds its way back to the atmosphere.

In the case of animals, whose diet is wholly vegetable, most of the nitrogen is eliminated as **hippuric acid**, $C_6H_5CO.NHCH_2COOH$. These two main end products of animal metabolism, viz., **urea** and **hippuric acid**, are not directly available for plant food. They must first be converted into ammonia, NH_3, by means of certain organisms contained in the soil. The most prominent of these organisms are the *Micrococcus ureæ* and the *Bacillus ureæ*. These organisms, being widely distributed, soon cause urine when exposed to the air to evolve ammonia. They are not contained in freshly excreted urine. The following is the change which takes place :—

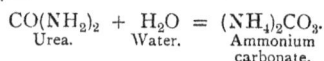

$$\underset{\text{Urea.}}{CO(NH_2)_2} + \underset{\text{Water.}}{H_2O} = \underset{\substack{\text{Ammonium} \\ \text{carbonate.}}}{(NH_4)_2CO_3}.$$

It is supposed that the change is brought about by an "enzyme" called **urease**, secreted by the micro-organisms.

The ammoniacal fermentation of the urea proceeds until about 13 per cent. of ammoniacal carbonate is formed in solution, when it ceases. Nitrogen in the form of ammonium carbonate is directly assimilable as a plant food, being built up again into vegetable proteins, which form the food of animals.

The hippuric acid is similarly decomposed according to the equation—

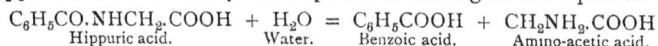

$$\underset{\text{Hippuric acid.}}{C_6H_5CO.NHCH_2.COOH} + \underset{\text{Water.}}{H_2O} = \underset{\text{Benzoic acid.}}{C_6H_5COOH} + \underset{\text{Amino-acetic acid.}}{CH_2NH_2.COOH}$$

Plants, however, not only absorb their nitrogen in the form of ammonia, but also in the form of oxidation products of ammonia, viz., nitrites and nitrates.

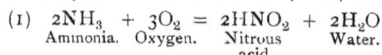

$$(1) \quad \underset{\text{Ammonia.}}{2NH_3} + \underset{\text{Oxygen.}}{3O_2} = \underset{\substack{\text{Nitrous} \\ \text{acid.}}}{2HNO_2} + \underset{\text{Water.}}{2H_2O}$$

$$(2) \quad \underset{\substack{\text{Nitrous} \\ \text{acid.}}}{HNO_2} + O = \underset{\substack{\text{Nitric} \\ \text{acid.}}}{HNO_3}$$

The process of oxidation of ammonia into nitrous and nitric acid is known as **nitrification,** and occurs in two distinct stages by means of two separate lots of organisms. The one kind of organism oxidises the ammonia to **nitrite,** and the other kind oxidises the nitrites to **nitrates.**

According to Boullanger and Massol there are two well-defined organisms which convert the ammonia into nitrites, viz., *Nitrosococcus*, a large, nearly spherical, organism, existing in two varieties, one found in Europe and the other in certain soils of Java. The other smaller organism is called *Nitrosomonas.*

The **nitric** organism is a very small bacterium, slightly longer than broad. The nitrous and nitric organisms exist side by side in nature, and neither can perform its work without the aid of the other, *e.g.*, the nitrous organism alone cannot carry the oxidation of the ammonia further than the stage of nitrite, while the nitric organism is incapable of directly oxidising ammonia.

In addition to the oxidising of ammonia to nitrites and nitrates (**nitrification**), there also goes on a reverse process (**denitrification**) whereby nitrites and nitrates are *destroyed* by the action of certain organisms, *e.g.*, a nitrate is converted directly into nitrogen, the final result being shown by some such end equation such as this :—

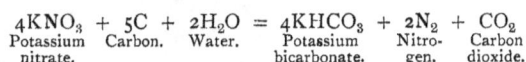

$$4KNO_3 + 5C + 2H_2O = 4KHCO_3 + 2N_2 + CO_2$$

| Potassium nitrate. | Carbon. | Water. | Potassium bicarbonate. | Nitrogen. | Carbon dioxide. |

The denitrifying organisms may be divided into two classes. (1) True denitrifying organisms, which actually destroy nitrates, converting them into free nitrogen. (2) Indirect denitrifying organisms, which reduce nitrates to nitrites, and these nitrites, when brought into contact with amido compounds such as urea, in the presence of acid, are decomposed with liberation of nitrogen, thus :—

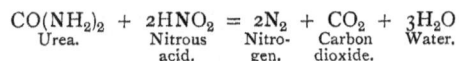

$$CO(NH_2)_2 + 2HNO_2 = 2N_2 + CO_2 + 3H_2O$$

| Urea. | Nitrous acid. | Nitrogen. | Carbon dioxide. | Water. |

By means of these changes much nitrogen from the animal or vegetable body is again thrown back into the atmosphere in its original form, thus maintaining the nitrogen cycle in nature.

The chief nitrogenous artificial manures used are ammonium salts and sodium or potassium nitrate. The quantities of these substances, however, at the disposal of agriculturists, is far less than the demand.

Until quite recently the supply of nitrates, and also of nitric acid, was derived almost exclusively from the natural deposits of **sodium nitrate,** $NaNO_3$, found in Chile (Chile saltpetre). The ammonium salts were principally derived from coal-tar and gas-making works, coming into the market as ammonium sulphate. The high price of these products, and the urgent need for more nitrogenous manure for agriculturists, has caused the successful invention recently of several processes for combining atmospheric nitrogen artificially, in a form assimilable by plants. The economic "fixation of nitrogen" has been achieved by three distinct methods which we will later deal with in detail, viz. :—(1) Direct oxidation of atmospheric nitrogen to nitric acid or nitrates. (2) Fixation of atmospheric nitrogen by means of calcium carbide, with formation of calcium cyanamide. (3) Direct formation of ammonia, NH_3, by direct combination of atmospheric nitrogen and hydrogen.

Other nitrogenous compounds much used are cyanides, ferro- and ferricyanides, etc. Cyanides find their chief use in the extraction of gold from quartz.

CHAPTER II

The Nitrate Industry

CHAPTER II

THE NITRATE INDUSTRY

LITERATURE

NORTON.—"Consular Report on the Utilisation of Atmospheric Nitrogen." Washington, 1912.
NEWTON.—*Journ. Soc. Chem. Ind.*, 1900, 408.
STURZER.—" Nitrate of Soda." 1887.
BILLINGHURST.—*Chem. Zeit.*, **11**, 752.
SIR WILLIAM CROOKES.—"The Wheat Problem."
THIELE.—"Saltpeterwirtschaft und Saltpeterpolitik." 1905.

SODIUM NITRATE (CHILE SALTPETRE), NaNO$_3$

SODIUM nitrate occurs in enormous quantities in the valleys of the Tarapaca and Tacoma, along the coast of Chile and Peru. Deposits also occur on the coast of Bolivia. The whole district is now a desert (rain only falling once in three years or so), but signs of former fertility are evident.

The main nitrate bed of Tarapaca is 2½ miles wide, and about 260 miles long, stretching along the eastern slope of the Cordilleras and the sea. The average distance from the sea coast is 14 miles, although some rich deposits occur inland as far as 90 miles. The deposits are some 500-600 ft. higher than the valley of the Tamaragal, the deposits thinning out as the valley is approached, all nitrate vanishing at the bottom.

The nitrate (locally called "caliche") forms a rock-like mass 3-6 ft. deep; it is covered over to a depth of 6-10 ft. by rocky matter, which in its turn is covered to a depth of 8-10 in. by fine loose sand. The nitrate lies upon a bed of clay, free from nitrate, which covers primitive rock. The colour of the nitrate varies from pure white to brown or gray.

The crude caliche contains from 10-80 per cent. of NaNO$_3$, much NaCl, a little KNO$_3$, some sodium perchlorate, NaClO$_4$, and sodium iodate, NIO$_3$.

A caliche containing 40-80 per cent. NaNO$_3$ is regarded as of the best quality; 25-40 per cent. NaNO$_3$ is second quality caliche, while anything below 25 per cent. NaNO$_3$ is regarded as inferior quality.

The richest beds, it is stated, have been practically worked out, but it is asserted that sufficient caliche still exists in Northern Chile and the neighbouring lands of Peru and Bolivia to supply the world's markets for another 100 years at the present rate of consumption.

The caliche is worked for refined nitrate on the places of production in about one hundred and fifty factories. The ground is broken by blasting, and the big blocks are broken into smaller ones by means of crowbars, and the lumps of caliche are separated from the rocky over-layer by hand. The caliche is crushed into 2-in. lumps, placed in long iron tanks heated with steam coils, and is there boiled with water. The earthy residue is left on a false bottom about 1 ft. from the real bottom of the tank, while the mother liquors, containing the NaNO$_3$ in solution, are run into the next tank. The extraction of the caliche is carried out systematically. The tank containing nearly exhausted caliche is treated with fresh water, while the liquor from this tank is run through a 9-in. pipe into the second tank, where it meets with a partly extracted caliche. From this, still rich in nitrate, the mother liquors pass on, finally emerging from the last tank of a battery of six filled with freshly charged caliche. The hot mother liquors from the last tank, at a temperature of 112°, and containing about 80 lbs. of sodium nitrate per cubic foot, are run off into the crystallising pans where, after cooling for four to five days, much nitrate is deposited in a crystalline condition.

The mother liquors now contain about 40 lbs. of nitrate per cubic foot, and are used in the systematic lixiviation of the caliche described above, being run (in addition to the weak liquor from the final exhaustion of the caliche) on to the subsequent caliche. These mother liquors, however, contain much iodine in the form of sodium iodate, from which the iodine must first be recovered in the manner described in the section on **Iodine** in Martin's " Industrial Chemistry," Vol. II. From the iodine house the mother liquors are passed on to the lixiviating tanks, to be used once more, as above described, in extracting crude caliche. The iodine recovered in this way is now an important article of commerce.

The crystals of nitrate deposited in the crystallising tanks are covered with a little water, drained, and allowed to dry in the sun for five or more days. They have then the following composition :—

	Best Quality.	Second Grade.
$NaNO_3$	96.5	95.2
$NaCl$	0.75	2.5
Na_2SO_4	0.43	0.7
H_2O	2.3	1.6

About 1 ton of coal is required for the production of 7 tons of nitrate.

This 95-96 per cent. $NaNO_3$ is shipped, and can be used directly as manure. About 20 per cent. of total export is used for the manufacture of nitric acid and other nitrogen compounds.

The second grade 95 per cent. $NaNO_3$ contains (as is evident from above analyses) much NaCl, and sometimes a little KNO_3 and sodium perchlorate, $NaClO_4$. This latter substance is especially undesirable in manurial nitrate, as it acts as a poison for plants. The nitrate is usually valued on the basis of its nitrogen content, the pure nitrate containing 16.47 per cent. N against 13.87 per cent. N in potassium nitrate (KNO_3, saltpetre).

Properties of Sodium Nitrate.—The substance forms colourless transparent anhydrous rhombohedra, whose form closely approximates to cubes (hence the term "cubic nitre"). It fuses at 316°, and at higher temperatures evolves oxygen with formation of nitrite. At very high temperatures all the nitrogen is evolved, and a residue of Na_2O and Na_2O_2 is left. The salt, *when quite free from chlorides of calcium and magnesium*, is not hygroscopic; as usually obtained, however, it is slightly hygroscopic. One hundred parts of water dissolve :—

Temperature	0° C.	10°	15°	21°	29°	36°	51°	68°
$NaNO_3$	66.7	76.3	80.6	85.7	92.9	99.4	113.6	125.1

Statistics.—The shipments of sodium nitrate from Chile and Peru are rapidly increasing, as the following figures show :—

Year.	Tons (Metric).	Year.	Tons (Metric).
1850	25,000	1909	2,100,000
1870	150,000	1910	2,274,000
1890	1,000,000	1911	2,400,000
1900	1,400,000	1912	2,542,000
1908	1,746,000		

In 1912 the 2,542,000 tons of nitrate were divided among the following countries :—United Kingdom, 5.60 per cent. of total ; Germany, 33.30 per cent. ; France, 14.30 per cent. ; Belgium, 12.2 per cent. ; Holland, 5.90 per cent. ; Italy, 2 per cent. ; Austria-Hungary, 0.25 per cent. ; Spain and Portugal, 0.50 per cent. ; Sweden, 0.15 per cent. ; United States, 22.2 per cent. ; Japan and other countries, 3.60 per cent.

In view of the rapidly increasing demand for sodium nitrate, there is grave doubt as to the possibility of any great extension in the demand being met by Chile. The caliche beds are spread over great areas in a desert region, where fuel and water are expensive, and are not uniform in quality. The richer and more easily extracted nitrate, the first to be worked, is stated to be confined to the valleys of the Tarapaca and Tacoma, and, it is asserted, will not last much longer than 1923 at the present rate of increased extraction. 100-150 years is the limit assigned for the working out of the less valuable beds in the hands of the Government. However, it must be remembered that perfectly enormous supplies of inferior nitrate remain in the soil, now regarded as not worth working, and possibly with improved methods these supplies will last much longer than the estimated time. In fact, fresh estimates by competent scientists put off the date of failure of the caliche beds for another 150 years at least.

Uses of Sodium Nitrate.—The main use of sodium nitrate is for manurial purposes, no less than 80 per cent. of total product being used for this purpose in 1912. Of the remaining 20 per cent used for manufacturing nitrogen compounds, about 15 per cent. was used for making nitric acid, and the remaining 5 per cent. for manufacture of potassium nitrate (saltpetre) for use in the lead chamber process for making sulphuric acid, and for other purposes as well.

For the manufacture of old black gunpowder or for fireworks, the slightly hygroscopic nature of $NaNO_3$ renders it inapplicable as a substitute for potassium nitrate, KNO_3.

POTASSIUM NITRATE, KNO_3.

Potassium Nitrate, KNO_3, crystallises in large prisms, isomorphous with rhombic aragonite, and so is sometimes known as "prismatic" saltpetre. It is used principally for the production of black shooting powder and for fireworks.

However, since the production of the new smokeless powders, the manufacture of black powder has greatly decreased, and with it the use of potassium nitrate has also decreased.

Manufacture.—Germany alone produces 20,000 tons of KNO_3, and Great Britain imports annually 10,000 tons, while over 30,000 tons are produced in other lands, almost entirely by the so-called "conversion process," using as raw materials the German potassium chloride, KCl (mined at Stassfurt) and Chile saltpetre, sodium nitrate, $NaNO_3$. This process depends upon the fact that under certain conditions of temperature and pressure, solutions of potassium chloride, KCl, and Chile saltpetre, $NaNO_3$, when mixed, undergo a double decomposition, sodium chloride being deposited and potassium nitrate remaining in solution :—

$$KCl + NaNO_3 = NaCl + KNO_3.$$

This conversion takes place all the more readily because KNO_3 is much more soluble in hot and much less soluble in cold water than is sodium nitrate. 100 g. of water dissolve :—

	KNO_3.	$NaNO_3$.	NaCl.	KCl.
At 20° C.	32 g.	88 g.	36 g.	34 g.
,, 100° C.	246 g.	176 g.	39.6 g.	56.5 g.

The potassium nitrate thus obtained is exceedingly pure. In fact, that technically produced for the manufacture of black powder must not contain more than 0.1-0.05 per cent. NaCl.

The process is carried out as follows :—In a large iron pan, provided with a stirring apparatus and indirect steam heating arrangement, there is added 188 kg. potassium chloride (88 per cent., containing NaCl) and 180 kg. Chile saltpetre (95 per cent.) and 160 kg. of mother liquors from a previous operation. The solution is boiled by indirect steam. The amount of water present is not sufficient to dissolve all the difficult soluble NaCl present, whereas the readily soluble KNO_3 at once goes into solution. The liquid is filtered hot, and the residual salt, after covering with a little water, and draining, is sold as such. The hot mother liquors are allowed to cool, when "raw" KNO_3 crystallises out—still in a crude state, however, containing several per cent. of NaCl.

The crude KNO_3 is once more dissolved by hot water, and allowed to crystallise in copper pans provided with stirrers. "Refined" potassium nitrate is thus obtained, which, after passing through a centrifugal machine, washing with very little water, and drying, is almost quite free from NaCl. The various mother liquors resulting from these operations are evaporated and added to the mother liquors for the initial treatment of the KCl and $NaNO_3$ as above described.

There is thus but little loss, although the calcium and magnesium salts which gradually accumulate in the mother liquors must from time to time be removed by the cautious addition of sodium carbonate. Sometimes a considerable amount of sodium perchlorate, $NaClO_4$, and sodium iodate accumulate in these mother liquors, and occasionally they are worked up for perchlorate and iodine.

Potassium nitrate, KNO_3, is one of the oldest known salts of potassium, and is characterised by the fact that when mixed with oxidisable matter it gives rise to readily inflammable products, such as black powder and mixtures used in fireworks. KNO_3 is the final result of the oxidation of nitrogenous organic material, and so it is steadily produced in the soil from decaying organic matter by means of special bacteria (see p. 5). It thus occurs, together with other nitrates, as an efflorescence in the soil in tropical countries like Bengal, Egypt, Syria, Persia, Hungary, etc. In India, and especially Ceylon, considerable quantities, *e.g*, 20,000 tons per year, of potassium nitrate are obtained by the lixiviation of certain porous rocks, which yield from 2.5-8 per cent. of their weight of nitrate.

Whether derived from rocks or from efflorescences in the soil near stables, urinals, etc., the nitrate is invariably produced by the decay of nitrogenous organic material, first into ammonia, and later, by oxidation, into nitrous and nitric acids, which, in the presence of alkali, produce potassium nitrate.

At one time much potassium nitrate was produced in Europe by the Governments of the different countries for supplying themselves with the necessary nitrate for making gunpowder, the operation being carried out in "*saltpetre plantations.*"

For this purpose nitrogenous organic matter of animal or vegetable origin was allowed to putrefy by exposure to air in a dark place ; it was then mixed with lime, mortar, and wood ashes (containing salts of potassium and sodium) and heaped into low mounds. These were left exposed to the air, being moistened from time to time by urine and the drainage from dung-heaps.

After a couple of years the outer surface of the saltpetre earth was removed, and the nitrates extracted by lixiviation with water. To the solution potassium carbonate was added, and on concentration and filtering from the precipitated calcium and magnesium salts, the clear solution was evaporated for KNO_3.

This industry, however, since the introduction of the cheap $NaNO_3$, has almost entirely ceased, the KNO_3 being now made from the $NaNO_3$ by the conversion process described above. It is possible that, in a modified form, this old industry may revive.

Thus Muntz and Laine (*Compt. rend.*, 1905, **141**, 861 ; 1906, **142**, 430, 1239) impregnated peat with sufficient lime to combine with the nitric acid formed, and then inoculated it with nitrifying bacteria and passed through it a 0.75 per cent. solution of ammonium sulphate $(NH_4)_2SO_4$, at 30° C., thereby obtaining a 1 per cent. solution of calcium nitrate, $Ca(NO_3)_2$. The bacteria would oxidise quickly only *dilute* solution of ammonium salts, but even **22** per cent. nitrate in the solution did not interfere with the process. Consequently, by sending the ammonium sulphate solution five times through the peat beds there was finally obtained a solution containing 41.7 g. of $Ca(NO_3)_2$ per litre.

Yield.—6.5 kg. of $Ca(NO_3)_2$ in twenty-four hours per cubic metre of peat. The old saltpetre plantations yielded 5 kg. KNO_3 in two years per cubic metre.

Properties.—White soluble crystals which readily dissolve in water, producing a great lowering of temperature, a fact which was once utilised for making freezing mixtures. 100 parts of water dissolved :—

Temperature -	0° C.	10°	20°	40°	60°	80°	100°	114.1°
KNO_3 - -	13.3	20.9	31.6	63.9	109.9	169	246	311

The saturated solution boils at 114.1° C. When heated KNO_3 evolves oxygen, and so is a powerful oxidising agent.

Uses.—Chiefly for the manufacture of black powder and fireworks, but also for pickling or salting meat, to which it imparts a red colour. It is also used in medicine and in the laboratory.

Calcium nitrate, $Ca(NO_3)_2 + 4H_2O$, containing 11.86 per cent. N, is now made by allowing nitric acid (made by the electrical oxidation of the atmosphere) to act on limestone (see p. 27). At the present time it is being produced in rapidly increasing quantities, and is now competing with Chile saltpetre (sodium nitrate) for manurial purposes. One great disadvantage to its use is its very hygroscopic nature; it is too readily washed from soils without complete assimilation of the nitrogen. It is now produced in Norway on a large scale; it is put on the market in coarse lumps, in a partly dehydrated condition, containing 13 per cent. N., being sold as "Norwegian saltpetre," or "nitrate of lime." It is largely used for preparing barium nitrate, $Ba(NO_3)_2$, and ammonium nitrate.

Ammonium nitrate, $(NH_4)_2NO_3$, is a product of increasing importance. It is used in large quantities for the manufacture of certain safety explosives, being especially suitable for this purpose because it leaves no solid residue on explosion, and develops heat when it decomposes according to the equation :—

$$2NH_4NO_3 = N_2 + O + 2H_2O.$$

also for making nitrous oxide :—

$$NH_4NO_3 = N_2O + 2H_2O.$$

Manufacture.—It may be produced by leading vapours of ammonia into nitric acid, using synthetic ammonia and Ostwald's nitric acid (see p. 29).

Regarding the production of ammonium nitrate by Ostwald's process of directly oxidising the ammonia to ammonium nitrate by catalytic platinum, the reader should see the following patents :— English Patents, 698 and 8,300, 1902 ; 7,908, 1908 ; American Patent, 858,904, 1907.
For a description of the process see Ostwald, *Berg. u. Hüttenm. Rundschan*, 1906, **3**, 71 ; see also Kaiser, English Patent, 20,305, 1910 ; see also Schmidt and Böcker, *Ber.*, 1906, p. 1366.
Frank and Caro (D.R.P., 224,329) propose to use thorium oxide as the catalyst. M. Wendriner (*Chem. Ind.*, 1911, p. 456) suggests uranium compounds as catalyst.
The Deutschen Ammoniak-Verkaufs-Vereinigung, in Bochum, produced by Ostwald's process in 1908, 651 tons ; 1909, 1,096 tons ; 1910, 1,237 tons.
In 1915 one firm alone in Germany estimate that they will produce 90,000 tons annually of this ammonium nitrate by this process.
Traube and Biltz (Swedish Patent, 8,944, 1897) oxidise ammonia gas by electrolytic oxygen, using as catalysers copper hydroxide, and claim an excellent yield.
Siemens and Halske (D.R.P., 85,103) oxidise ammonia gas by silent electrical discharges.
Nithack (D.R.P., 95,532) electrolyse water, containing nitrogen, dissolved under a pressure of 50-100 atmospheres.

The bulk of the ammonium nitrate at the time of writing is made, however, by treating calcium nitrate (made as above described) with rather less than the equivalent quantity of ammonium sulphate :—

$$Ca(NO_3)_2 + (NH_4)_2SO_4 = CaSO_4 + 2NH_4NO_3$$

(see Wedekind's patent, English Patent, 20,907, 1909).
The yield is quantitative, the $CaSO_4$ being removable without difficulty.
For a complete summary of patents see *Zeit. gesammt. Scheiss. Sprengstaffwesen*, 1914, **9**, 81.

Regarding the manufacture of ammonium nitrate from **calcium nitrate** the reader should consult the patents :—Dyes, English Patent, 15,391, 1908 ; Nydegger and Wedekend, English Patent, 20,907, 1909.
The Norsk Hydro-Elektrisk-Kraelstof-Aktieselshab (D.R.P.,206,949 of 1907) prepare ammonium nitrate by leading nitrous gases through calcium cyanamide :—

$$CaCN_2 + 2H_2O + 4HNO_3 = Ca(NO_3)_2 + CO_2 + 2NH_4NO_3.$$

See also the French Patent, 417,505 of 1910 ; English Patent, 19,141 of 1910 ; Colson, *Journ. Soc. Chem. Ind.*, 1910, p. 189 ; Garroway, English Patent, 7,066, 1897.
Preparation of Ammonium Nitrate by the Ammonia Soda Process.—In 1875 Gerlach (English Patent, 2,174) attempted to produce ammonium nitrate from ammonium carbonate and Chile saltpetre according to the equation :—

$$NH_4HCO_3 + NaNO_3 \rightleftharpoons NaHCO_3 + NH_4NO_3.$$

The process, however, failed on account of the incomplete decomposition (see *Journ. Soc. Chem. Ind.*, 1889, p. 706).

The following patents bear on this :—Lesage & Co. (French Patent of 20th Jan. 1877), Chance (English Patent, 5,919, 1885), Fairley (D.R.P., 97,400; English Patent, 1,686, 1896), *cf.* Lunge (" Sulphuric Acid and Alkali," Vol. III.), Feld (English Patent, 5,776, 1906 ; American Patent, 839,741 ; D.R.P., 171,172, 178,620) have recently improved the process.

The theory of the process, according to the Phase Law, has been worked out in full by Fedotieff and Koltunov, *Zeits. anorg. Chem.*, 1914, **85**, 247-260.

Numerous attempts have been made to realise the production of ammonium nitrate from the equation :—

$$2NaNO_3 + (NH_4)_2SO_4 = Na_2SO_4 + 2NH_4NO_3.$$

Concerning this the following patents should be consulted :—Roth (D.R.P., 48,705, 53,364, 55,155, 149,026), F. Benker (D.R.P., 69,148), T. Fairley (English Patent, 1,667, 1896), J. V. Skoglund (D.R.P., 127,187, 1901), E. Naumann (D.R.P., 166,746), Caspari, Nydegger, and Goldschmidt (American Patent, 864,513 ; D.R.P., 184,144), C. Craig (English Patent, 5.815, 1896 ; D.R.P., 92,172), R. Lennox (D.R.P., 96,689), Wahlenberg (English Patent, 12,451, 1889), Campion and Tenison-Woods (English Patent, 15,726, 1890).

Ammonium nitrate is extremely soluble in water and hygroscopic. These properties mitigate to some extent against its usefulness as a manure. When dissolved in water a lowering of temperature up to $-16.7°$ C. is attainable. Owing to its great solubility in water it cannot be manufactured by the double decomposition of sodium nitrate with ammonium salts.

Sodium nitrite, $NaNO_2$. Some 5,000 tons of this substance are now annually produced for the purpose of diazotisation in the manufacture of dyes (see Martin's " Industrial Chemistry," Vol. I.).

Until recently it was made solely from Chile saltpetre by heating with metallic lead to 450° C. :—

$$NaNO_3 + Pb = NaNO_2 + PbO.$$

See J. Turner, *Journ. Soc. Chem. Ind.*, 1915, **34**, 585, for details.

At the present time it is now almost exclusively produced by the cheaper process of leading the *hot* nitrous gases coming from the furnaces for the manufacture of nitric acid by the Birkeland-Eyde process (pp. 27, 28) up a tower down which streams $NaOH$ or Na_2CO_3 solution.

When the hot nitrous gases from the furnace are thus treated when at a temperature of 200°-300° C. (when they consist of a mixture of NO and NO_2 molecules), there is produced a nitrate-free sodium nitrite, $NaNO_2$, which is now manufactured in this way in Christiansand. For further particulars see p. 28.

CHAPTER III

The Nitric Acid Industry

CHAPTER III

THE NITRIC ACID INDUSTRY

LITERATURE

A. W. CROSSLEY.—"The Utilisation of Atmospheric Nitrogen" (Thorpe's "Dictionary of Applied Chemistry," Vol. III., p. 698). 1912. (An excellent account.)

T. E. NORTON.—"Consular Report on the Utilisation of Atmospheric Nitrogen." Washington, 1912. (Gives statistics.)

J. KNOX.—"The Fixation of Atmospheric Nitrogen." 1914. (Gives much of the literature.)

DONATH AND INDRA.—"Die Oxydation des Ammoniaks zu Salpetersäure und Salpetriger Säure." 1913.

DONATH AND FRENZEL.—"Die technische Ausnutzung des atmosphärischen Stickstoff." 1907.

ESCARD.—"Fabrication électrochemique de l'Acide Nitrigue et des Composés Nitrés à l'Aide des Eléments de l'Air." Paris, 1909.

CARO.—*Chemical Trade Journal*, 1909, **44**, 621.

HABER AND KOENIG.—*Zeitsch. Elektrochem.*, 1910, **16**, 17.

E. KILBURN SCOTT.—"Production of Nitrates from the Air," *Journ. Soc. Chem. Ind.*, 1915, **34**, 113.

See also numerous references in text.

NITRIC ACID, HNO$_3$

WITHIN the last thirty to forty years a very great change has come over the character of the nitric acid manufactured. At one time the only acid placed on the market was an acid of 55-68 per cent. HNO$_3$ (1.35-1.41 sp. gr.), and the somewhat stronger "red fuming" acid; at the present time the bulk of the nitric acid made has a strength of 95 per cent. and over (1.5-1.52 sp. gr.). In order to produce an acid of this strength very considerable difficulties have to be overcome, since 100 per cent. HNO$_3$ is only stable below 0° C., and begins to boil at 86° C., evolving red fumes and partially decomposing according to the equation :—

$$2HNO_3 = NO + NO_2 + O_2 + H_2O.$$

As the nitrous gases escape the boiling point of the acid rises, and it is only when a liquid containing 65 per cent. HNO$_3$ (1.4 sp. gr.) is attained that the mixture distils over unchanged under atmospheric pressure at 120° C. These difficulties are now almost completely overcome with modern apparatus.

Nitric acid is now made on the large scale by three distinct processes :—

(1) *Manufacture from Chile Saltpetre and Sulphuric Acid.*
(2) *Manufacture by Electrical Oxidation of the Atmosphere.*
(3) *Manufacture by Catalytic Oxidation of Ammonia.*

We will consider each process in detail.

(1) **Manufacture of Nitric Acid from Chile Saltpetre.**—Nearly equal weights of Chile saltpetre, NaNO$_3$, and "oil of vitriol" are heated together in large iron retorts, often capable of holding some 5 tons or more of the mixture. The quantities taken correspond to the equation :—

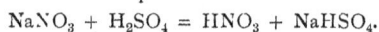

$$NaNO_3 + H_2SO_4 = HNO_3 + NaHSO_4.$$

2

This reaction is carried to completion at a temperature under 150° C., thereby avoiding much loss by decomposition of the nitric acid. A much higher temperature with half the amount of sulphuric acid is necessary to cause the formation of the neutral sulphate, according to the equation :—

$$2NaNO_3 + H_2SO_4 = Na_2SO_4 + 2HNO_3.$$

This latter action, however, in the best practice is never used, because the high temperature necessary to complete the action causes a very considerable decomposition of the nitric acid formed ; also the neutral sodium sulphate remaining sets to a solid mass, and causes great difficulty as regards removal from the retort ; also the action takes a much longer time to complete, and requires more fuel.

By working not much over 150° C. and using more sulphuric acid, as corresponds to the first equation, the "bisulphate" is obtained as an easily fusible mass, which is withdrawn in a fluid condition from the retort merely by opening a plug at the base. However, in actual practice,

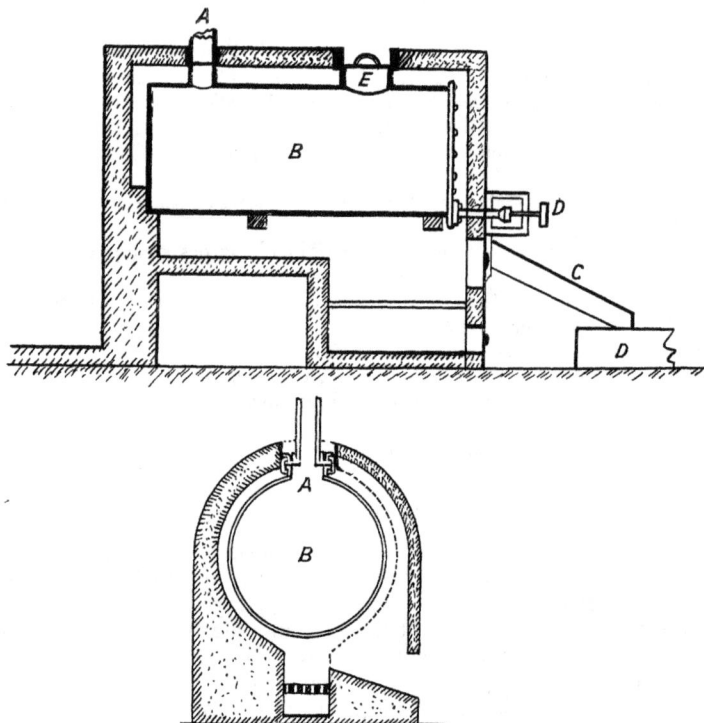

FIG. 1.—Nitric Acid Still.

right towards the end of the action the temperature is raised somewhat, when the bisulphate foams somewhat and is probably converted into pyrosulphate, thus :—

$$2NaHSO_4 = Na_2S_2O_7 + H_2O.$$

This pyrosulphate (or disulphate) is run off in a fluid condition into iron pans, where it solidifies and is used for the manufacture of normal sodium sulphate by heating with salt, NaCl, in mechanical sulphate furnaces. Part of this "**nitre cake,**" as the product is called, is now converted into sodium sulphide, Na_2S.

The kind of sulphuric acid used varies with the nature of the nitric acid it is desired to produce. For a dilute nitric acid of 1.35-1.4 sp. gr. it is sufficient to use 60° Bé. (141° Tw.) sulphuric acid direct from the Glover tower. The first distillate will be strongest, and the last nearly pure water. Should, however, a stronger HNO_3 of over 1.5 sp. gr. be required (as is now nearly always the case), a more concentrated 66° Bé. (167° Tw.) sulphuric acid must be employed, and the $NaNO_3$ is often previously melted.

Excess of H_2SO_4 is bad, for although it does not injure the quality of the HNO_3 produced, yet it diminishes the value of the $NaHSO_4$ or $Na_2S_2O_7$ (**nitre cake**) produced as a by-product for the manufacture of sodium sulphate.

We have seen above that when nitric acid boils, a mixture of NO and NO_2 gases is produced, and the NO_2 condensing, would cause the nitric acid produced to be contaminated with nitrous acid $(2NO_2 + H_2O = HNO_2 + HNO_3)$. This contamination may be very simply avoided by keeping the condensed distillate hot until it escapes from the cooling worm, or else by blowing air through the hot acid by means of an injector.

The plant used in practice is shown in Fig. 1, which shows the usual cylindrical cast-iron boiler B, the sulphuric acid and nitrate being introduced at E, while the nitric acid distils away through A. Such a cylinder will hold over 5 tons of the mixture, and is heated over a free fire. The cast iron is scarcely attacked by the nitric acid vapour.

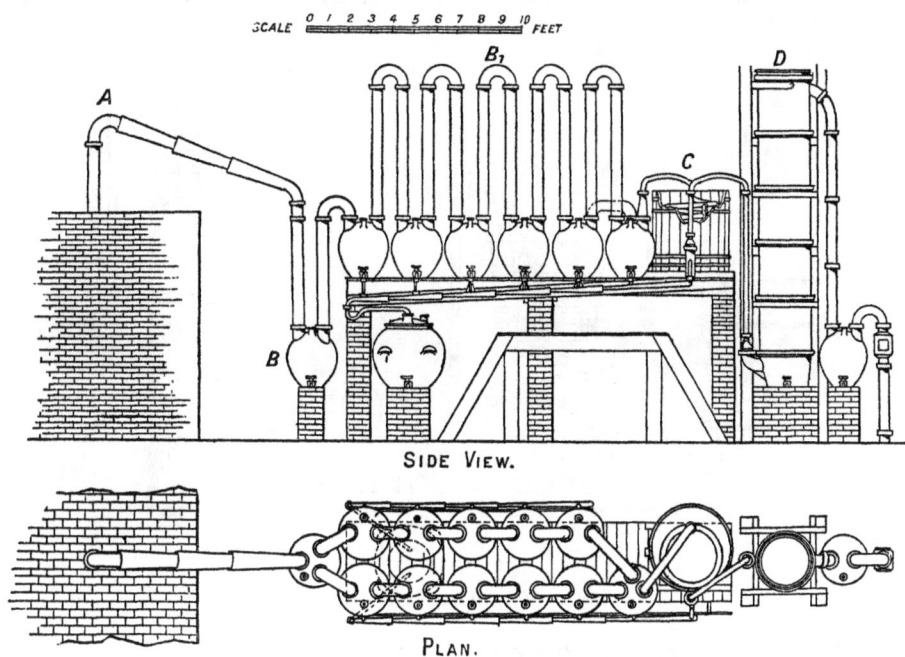

FIG. 2.—Doulton's Nitric Acid Condensing Plant.

The retort is emptied after the distillation by turning the screw stopper D, when the fused bisulphate and pyrosulphate pours away through the trough C into an iron pan E.

For very small installations sometimes an earthenware "pot" retort is used.

Since the specific heat of strong nitric acid is small, the fuel consumption for distillation is low ; the condensation of the vapours is readily effected.

The condensing arrangement consists of a series of earthenware or stoneware vessels, pipes, etc., which are luted together by means of an asbestos water-glass lute.

Fig. 2 shows an air-cooled condensing arrangement made by Doulton & Co., of Lambeth, London. A is the still head, the nitric acid from which passes into a preliminary stoneware receiving vessel B, and thence into a series of air-cooled vessels B_1. C is an earthenware condensing worm immersed in a cooling tank of water, while D are condensing towers, usually packed with glass balls or cylinders, or sometimes provided with plate columns, down which trickles water to absorb the last trace of free acid.

Fig. 3 shows a "Griesheim" condensing system, much used in Germany. Here the nitric acid, distilling out of retort A, passes first into B and then into an earthenware condensing coil E, standing in cold water. The vapours then pass through the pipe F, through a series of earthenware receivers, and then into the plate column condensing tower G.

First of all, at the lowest temperature the strongest acid distils over, contaminated, however, by HCl, NO_2Cl, and similar chlorine-containing volatile compounds. Next, there distils over the main quantity of strong acid, fairly pure, but contaminated with red nitrous fumes. To oxidise these a stream of hot air is blown in at K.

In some forms of apparatus (*e.g.*, Oscar Guttmann's) the oxidation is automatically effected by causing the gases coming from the still to pass into an enlargement in the pipe, where they produce an injector action and suck in a quantity of air through perforations in the socket pipe, thus causing oxidation of the nitrous to nitric acid.

The air not only purifies the condensed nitric acid from nitrous acid, but also frees it from volatile chlorine compounds, such as NO_2Cl. By regulating the temperature of the water in which the cooling coil E (Fig. 3) is immersed, the quality of the acid collecting in B can be also regulated to some extent, and the acid can be withdrawn in successive fractions as it comes over into the receiver C. In the tower G (Fig. 3) all the nitrous fumes finally escaping condensation are absorbed by the down-streaming water, so as to form a dilute nitric acid of 1.35-1.36 sp. gr. The water enters at *h*.

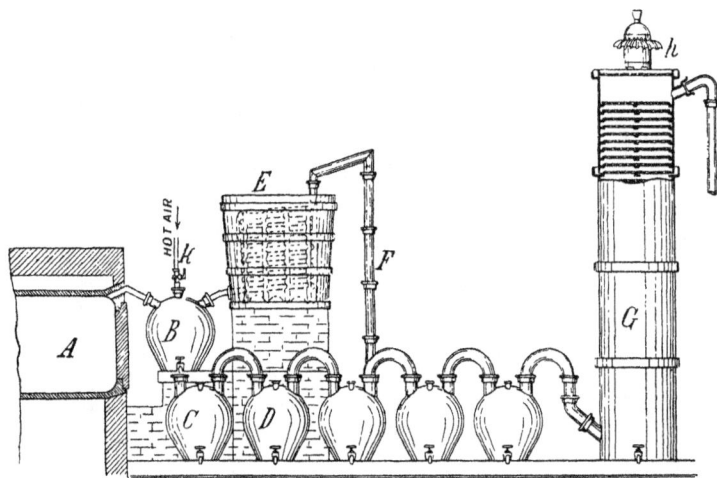

FIG. 3 —The Griesheim Nitric Acid Plant.

Other excellent condensing systems have been devised by Hart, Thomson, and others.

Valentiner's Vacuum Process.—In this process the outlet pipe from the still passes into a small receiver, half filled with pumice stone, so that any acid or solid coming over as spray or foam from the still is retained. The vapours then pass on through two water-cooled stoneware coils arranged in series; the vapour is next passed through two separate series of receivers.

Here, by means of a three-way tap, the condensed acid may, either at the beginning or end of the distillation, be collected in a large receiver while the concentrated 96 per cent. HNO_3, coming over intermediately, may be collected in the separate receivers. From these receivers the vapours pass into a reflux condenser consisting of an earthenware coil immersed in water.

Next, the vapours pass through eight receivers, alternately empty and half-filled with water. Finally come seven receivers, alternately empty and half-filled with milk of lime, which removes the last traces of the nitrous gases. The last receiver is connected with a vacuum pump.

This system offers great advantages, since the decomposition of the strong acid is largely avoided under the diminished temperature of distillation, resulting from the diminished air pressure over the still. The time of distillation is also greatly reduced, thereby causing a great saving in fuel. The process is gaining ground, since no leakage of injurious fumes and a larger and purer yield of acid is claimed. The first cost, however, is heavy (see United States Patent, 920,224).

(2) **Manufacture of Nitric Acid from the Atmosphere.**—The atmosphere contains about 4,041,200,000,000,000 tons of nitrogen, together with about one-fifth this amount of oxygen gas. There is thus an inexhaustible supply of nitrogen available for conversion into nitric acid or nitrate, and thus available for manurial purposes. The nitrogen in the air over a single square mile of land amounts to nearly 20,000,000 tons of nitrogen—sufficient, if suitably combined as nitrate or nitric acid, to supply at the present rate of supply the world with manure for over half a century.

The problem of directly uniting the oxygen and nitrogen of the air to form nitric acid and nitrate was first solved in a commercially successful form in 1903 by Birkeland and Eyde. In 1907 they established a factory at Notodden, in Norway, utilising the 40,000 H.P. there available from waterfalls. In the same year the Badische Anilin- und Soda-Fabrik built an experimental works at Christiansand, in Norway, and an agreement was come to between the interested firms whereby a company possessing a total capital of 34,000,000 kronen came into existence, which in 1911 opened a large works on the Rjukan Falls, near to Notodden, with 125,000 H.P. However, the Badische Anilin- und Soda-Fabrik have since withdrawn all their capital from the concern, selling all their rights in the manufacture of nitric acid from the atmosphere to the French Scandinavian Co. The cause of this is possibly the perfection by the B.A.S.F. of their process for making synthetic ammonia cheaply and thence oxidising it into nitric acid (see p. 28). Other works are building or are built in Norway and Sweden, and similar factories, using different systems as described below, have come into existence at Innsbruck, in Italy, France, the United States and other countries possessing considerable water power for the generation of electrical power. Altogether the industry is now well established.

At the present time, however, 1 H.P. per year can only produce $\frac{1}{2}$ to 1 ton of nitrate per year. There are, it is calculated, some 5 to 6 million H.P. available in Norway, and some 50,000 to 60,000 H.P. available in Europe, so that were it possible to harness all this power for the sole manufacture of nitrates from the air it would be possible to produce some 50,000,000 tons of nitrate annually. The amount now produced is far under 200,000 tons annually.

The process adopted for making nitric acid from the air is to oxidise the latter by means of an electric arc, whereby the nitrogen and oxygen directly unite to form nitric oxide, NO. This latter takes up oxygen from the air to form nitrogen peroxide, and the nitrogen peroxide is then absorbed by water, with the production of nitric and nitrous acid.

Theoretical.—When air is strongly heated, atmospheric oxygen and nitrogen directly unite to form nitric oxide, NO, according to the equation—

$$N_2 + O_2 = 2NO - 43,200 \text{ calories.}$$

The process is reversible, and depends upon the equilibrium represented by the expression :—

$$K = \frac{P_{(O_2)} P_{(N_2)}}{P^2_{(NO)}},$$

where P represents the partial pressures of the respective gases, and K is the reaction constant.

Also the reaction is endo-thermic (*i.e.*, the formation of NO is attended with the absorption of heat), and consequently is favoured by a high temperature, as is shown by the following results of Nernst (*Zeit. anorg. Chem.*, 1906, **49**, 213 ; *Zeit. Elektrochem.*, 1906, **12**, 527) interpolated for simplicity :—

1. Temperature °C.	2. Per Cent. Volume NO Observed.	3. Per Cent. Volume NO Calculated.
1,500	0.10	...
1,538	0.37	0.35
1,604	0.42	0.43
1,760	0.64	0 67
1,922	0.97	0.98
2,000	1.20	...
2,307	2 05	2.02
2,402	2.23	2.35
2,500	2.60	...
2,927	5.0 (approx.)	4.39
3,000	5.3	...

The second column represents the volume of NO found in 100 volumes of air, which have been exposed to the temperature indicated in the first column. The third column represents the amount of NO which should have been found as calculated from the law of mass action and the heat of formation.

It will be seen from this that the reversible reaction, $N_2 + O_2 \rightleftharpoons 2NO$, is very incomplete, only a small but definite amount of NO being produced, the quantity, however, *increasing as the temperature rises*, a very large quantity of NO being found above 3,000° C. Now Nernst and Jellinek showed that the equilibrium, $N_2 + O_2 \rightleftharpoons 2NO$, is almost *immediately established* (in a fraction of a second) at temperatures over 2,500°, but below this an appreciable time is required to establish equilibrium, **hours** being required below 1,500°.

This is clear from the following table, compiled from Jellinek's results (*Zeit. anorg. Chem.*, 1906, **49**, 229) :—

Absolute Temperature.	Time of Formation of Half the Possible Concentration of NO from Air.
1,000°	81.62 years
1,500°	1.26 days
1,900°	2.08 mins.
2,100°	5.06 secs.
2,500°	1.06×10^{-2} secs.
2,900°	3.45×10^{-5} secs.

Consequently, the decomposition of NO into N_2 and O_2 proceeds so slowly that its isolation is possible in appreciable quantities, *if the cooling to below 1,500° C. is carried out rapidly enough.*

To make this clear to the technical student, we will suppose that (according to Nernst results above tabulated) air is exposed to a temperature of 3,000° C. Then in 100 vols. of this air will be found 5.3 vols. of NO, and this equilibrium will be established in the fraction of a second. Let now this 100 vols. at 3,000° C. of air be suddenly cooled to say 1,500° C. Then, according to the equilibrium conditions tabulated by Nernst in the above table, the 5 per cent. of NO in the 100 vols. of air will gradually decrease, according to the equation, $2NO \rightleftharpoons N_2 + O_2$ until, after waiting until complete equilibrium is attained at 1,500° C., only 0.1 per cent. vol. of NO is left, as corresponds to equilibrium at this temperature of 1,500° C. However, to reach this equilibrium at the lower temperature requires some time, and if the cooling takes place in a fraction of a second, practically **all** the 5.3 per cent. of NO in the air at 3,000° will be found in the air at 1,500° (as the NO will not have had time to decompose), and if the air be kept for many hours at 1,500° the percentage of NO will gradually decrease from 5 per cent. in the sample of air until it reaches that which corresponds to *true* equilibrium at 1,500° C., viz., 0.1 per cent. NO. But now suppose we do not stop the cooling process at 1,500°, but cool it in a fraction of a second from 3,000° C. down to 1,000° C., then practically we will have 5 per cent. NO in our sample of air at 1,000° C., and although the equilibrium at this temperature requires that practically all the NO would decompose into O and N, thus :—$2NO \longrightarrow N_2 + O_2$, yet we would have to wait years for this action to proceed to completion, and so we could keep our specimen of air containing 5 per cent. NO for many hours at 1,000° C. without much loss.

Hence we may formulate the conditions of practical success for the isolation of NO as follows :—

NO is endothermic, and like other endothermic substances its stability **increases** as the temperature rises, and it is capable of existing in very large quantities at very high temperatures, such as 3,000°-5,000° C. It is also capable of existing stably **below** 1,000° C., but is unstable at intermediate temperatures. Hence in order to isolate reasonable quantities of NO we must (1) cause the formation of NO to take place at very high temperatures, such as 3,000°-10,000 C., when large amounts are formed ; (2) then extremely rapidly cool the NO formed through the intermediate unstable range of temperature, 2,500°, to a stable low temperature, *e.g.*, 1,000° C. If the cooling is effected quickly enough, although some of the NO is decomposed as it cools through the unstable lower temperature, yet sufficient survives so as to make an appreciable amount still existing at 1,000° C., which then ceases to decompose further, and so is available for conversion into nitric acid. Now in practice the intense heating of the air to very high temperatures such as 3,000-5,000° C., necessary to cause a reasonable oxidation of the air to NO, is always produced by a high tension alternating current of some 5,000 volts.

This electric current not only produces the union of the O and N as a purely thermal effect, but also ionises the air and causes a direct union by means of electrical energy, much as ozone is formed from oxygen by means of a silent electrical discharge. So that much larger quantities of NO are obtainable than strictly corresponds to the thermal conditions, *e.g.*, Haber and König (*Zeit. Elektrochem.*, 1907, **13**, 725 ; 1908, p. 689), by means of a (relatively) cold electrical arc of only 3,000° C. was able to isolate no less than 14.5 vol. of NO by mixing equal volumes of nitrogen an i oxygen gases together. The union takes place in the proportions $N_2 : O_2$, and therefore the mixture of gases in the atmosphere, *e.g.*, 21 per cent. O to 78 per cent. N, is not the most favourable mixture for the formation of NO. Technically, however, the addition of O to the atmosphere to improve the yield of NO is inapplicable at present on account of cost.

For technical success it has been found essential to use a high-tension alternating arc. By employing a low-tension electrical arc (*e.g.*, such as is used for manufacture of calcium carbide or for ordinary arc lamps) or by employing electrical sparks from a Rühmkorff coil, only very small amounts of NO were obtainable.

Another essential condition for technical success is the sudden and rapid cooling of the enormously hot gas in the arc (*e.g.*, 3,00c°-10.000° C.) to a temperature below 1,500° C.

We will describe only the following three electrical furnaces, all of which have been used on a large scale with technical success for combining atmospheric oxygen and nitrogen.

(1) *The Birkeland-Eyde Furnace.*
(2) *The Pauling Furnace.*
(3) *The Schönherr Furnace.*

(1) **The Birkeland-Eyde Furnace**.—Fig. 4 is a diagrammatic sketch of this furnace. The electrodes consist of two copper pipes A and B, kept cool by a current of water, and separated by 8-10 mm. They are connected with a high-tension powerful alternating current of 5,000 volts, which forms an arc between them. The arc is placed between the poles of a powerful electromagnet, which then blows it out into a wheel-like disc of flame, 2 m. in diameter, composed of burning oxygen and nitrogen. The whole is enclosed in a refractory casing, shown in section in Fig. 5 and a general view in Fig. 6. The section, Fig. 5, shows how air is blown in to feed the flame. The air enters at A A and passes in

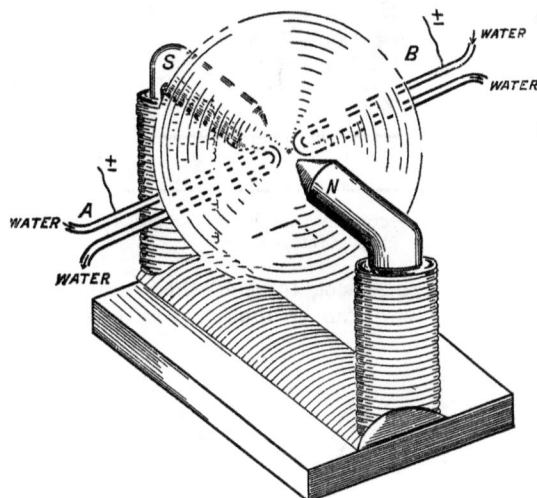

FIG. 4.—Diagrammatic View of the Birkeland-Eyde Electric Furnace.

through holes in the refractory lining. The electric flame plays down the disc-like space C C, and the burnt gases come out at D and then pass away to the absorption plant, as indicated in Fig. 5. E E are the wires of the electromagnets.

For further particulars the reader should see Crossley, *loc. cit.*; also the *Zeit. Elektrochem.*, 1905, **II**, 252; Birkeland, *Trans. Faraday Soc.*, 1906, **2**, 98; Eyde, *Journ. Roy. Soc. Arts*, 1909, **57**, 565; Scott, *Journ. Soc. Chem. Ind.*, 1915, **34**, 114.

In spite of the enormously high temperature of the disc flame (over 3,500° C.), the temperature of the walls do not rise above 800° C. owing to the cooling effect of the current of air. The temperature of the gases escaping from the furnace is 800°-1,000° C. They contain 1.5-2 per cent. NO. In 1907 no less than thirty-six furnaces were employed at Notodden in Norway, each taking up 800 kw. and 26,000-28,000 l. of air per minute pass through each furnace. The 40,000 H.P. employed is derived from water power. Still larger furnaces taking 3,000 kw. are now in use. The copper electrodes last 400-500 hours before they must be repaired.

According to Haber and Koenig (*Zeit. Elektrochem.*, 1910, **16**, 11) the Birkeland-Eyde furnaces yield 70 g. of HNO_3 per kilowatt hour, and the concentration of the NO produced is 2.0 per cent. It may be taken as approximately correct that 1 H.P. per year can produce about ½-1 ton of calcium nitrate.

(2) **The Pauling Furnace**.—This is shown in Fig. 7. The main electrodes

A and B, are bent into the shape of a **V**, A H H being a section of one main electrode, and B K K a section of the other.   The base of the electrodes thus forms

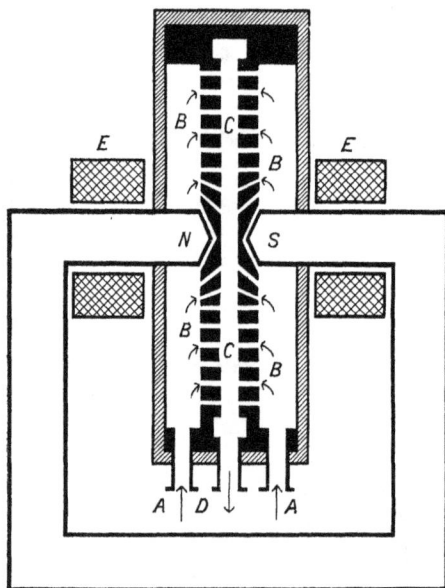

FIG. 5.—Section through the Birkeland-
Eyde Furnace.

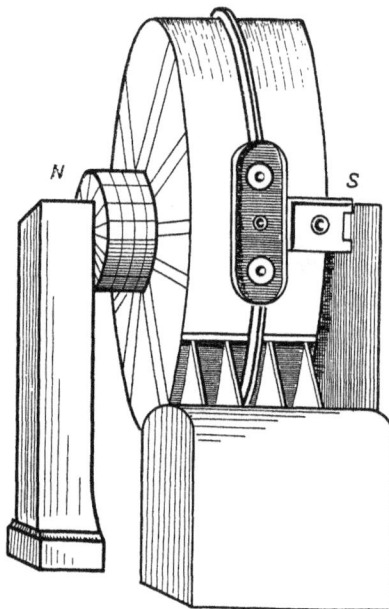

FIG. 6.—External View of the
Birkeland-Eyde Furnace.

at M N a vertical slot, through which are introduced thin "lighting knives F F." These "knives" can be brought very close together, 2-3 mm., by the screwing

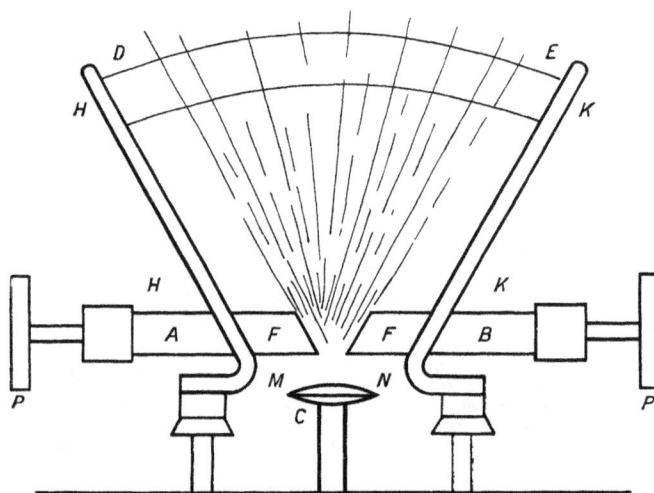

FIG. 7.—The Pauling Furnace.

arrangement P P, and the arc, thus lighted at the narrowest portion of the spark gap, shows a tendency to rise up between H H and K K, owing mainly to the upward pull of the hot gases, but is interrupted at every half period of the alternating current, only to be reformed at the lowest and narrowest part of the electrodes.

Through a nozzle c a stream of previously heated hot air is blown upwards into the arc, causing the air to diverge and form between the V-shaped main electrodes a flame of burning O and N, sometimes a metre in length. The very thin ignition blades soon burn away, requiring replacing every 20 hours. The main electrodes require replacing every 200 hours. Two such arcs in series are contained in each furnace, which is made of refractory material.

The gases leave the furnace at a temperature of 700°-800° C. containing 1.5 per cent. NO. Twenty-four of these furnaces, absorbing 15,000 H.P., were in 1911 at work at Patsch, near Innspruck, by the "Salpetersaure Industrie Gesellschaft," each furnace taking 400 kw. at 4,000 volts, and being supplied with 600 cm. of air per hour. Much larger furnaces of this type are now being erected at works near Milan (10,000 H.P.) and at Roche-de-Dame in South France (10,000 H.P.).

According to Haber and Koenig (*Zeit. Elektrochem.*, 1910, **16**, 11) the Pauling furnace produces 60 g. of HNO_3 per kilowatt hour, the escaping gases having a concentration of 1-1.5 per cent. NO.

For further particulars of this process see Scott, *Journ. Roy. Soc. Arts*, 1912, **60**, 645; also *Zeit. Elektrochem.*, 1907, **13**, 225; 1909, **15**, 544; 1911, **17**, 431.

(3) The Schönherr Furnace —Fig. 8

gives a view of this furnace. A A is an insulated high-tension iron electrode, the other electrode being the iron piping E E, into which A A projects, and an opposing electrode G_1. An arc is thus formed between the electrode A A and the iron piping; but a stream of air is blown in peripherically at the base of the piping, through a series of orifices X X, in such a way as to cause a whirling movement in the tube E F, and so causes a whirling flame of burning O and N to run up the tube E E E, being cooled at the top by the water-cooling arrangement F F. The hot nitrous gases stream away from E E, down the external pipes H H, and so out through M, into the plant for absorbing the nitrous fumes. The air enters the furnace at c, and is heated to a high temperature (about 500° C.) before being blown into the arc (through the orifices at x) by passing up the tubes D D, and down the tube B B, both of which are heated by the hot gases streaming away from the furnace. The arc is started by pressing the lever z, which brings an iron bar into momentary contact with the iron electrode A. The arc burns as steadily as a candle flame, and is observed through peepholes at N and O. The furnace is connected electrically to earth, so that all parts can be handled with impunity, except only the insulated electrode A. The electrode A requires replacing every four months.

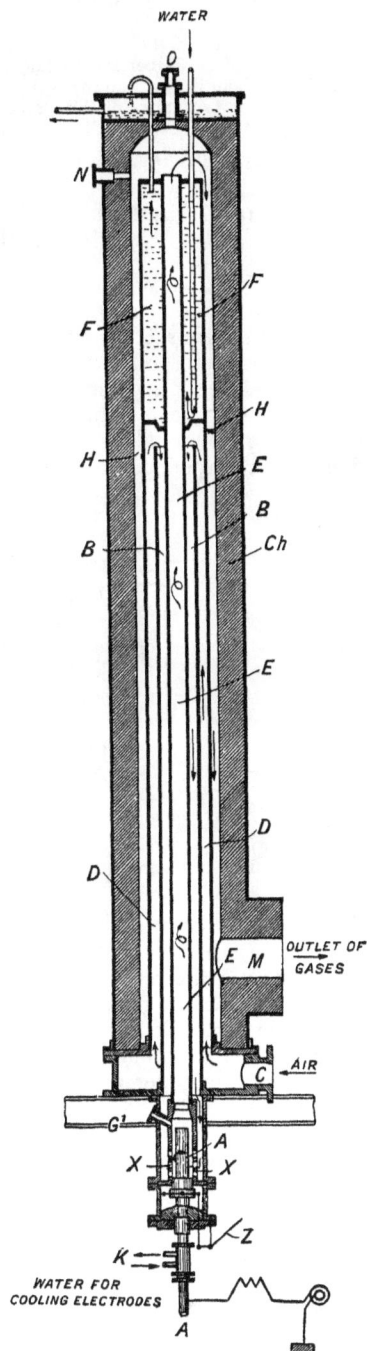

FIG. 8.—The Schönherr Furnace.

At Christiansand, in Norway, the Badische Anilin- und Soda Fabrik erected in 1907 a factory employing these furnaces, each being supplied with 600 H.P. at 4,200 volts. Others are projected using 750 H.P. and requiring 40,000 cub. ft. of air per hour, producing arcs 7 yds. long.

The gases leave the furnace at a temperature of 850° C., and contain 2-2.5 per cent. NO. According to Haber and Koenig, *Zeit. Elektrochem.*, 1910, **16**, 11, the yield in grams HNO₃ per kilowatt-hour is 75 g. and the NO produced amounts to 2.5 per cent.

For further particulars of this process see Schönherr, *Trans. Amer. Elektrochem. Soc.*, 1909, **16**, 131.

E. K. Scott (*Journ. Soc. Chem. Ind.*, 1915, **34**, 113) describes a new type of furnace.

General Plan of Plant for the Manufacture of Nitrates by Electrical Oxidation of the Air

How the manufacture of nitric acid and nitrates from the air is carried out technically by means of the electrical furnaces above described will be seen from the diagrammatic sketch, Fig. 9. A is an air compressor which drives the air into the electrical furnace B (which may be any of the kinds described above). In this chamber B, the air is passed into a very hot electrical arc flame, where it is heated to a very high temperature, say 5,000°-10,000° C., in the immediate path of the arc. Here union takes place, NO being formed. The hot gases passing away from the arc experience a sudden fall of temperature (although the fall is so slow

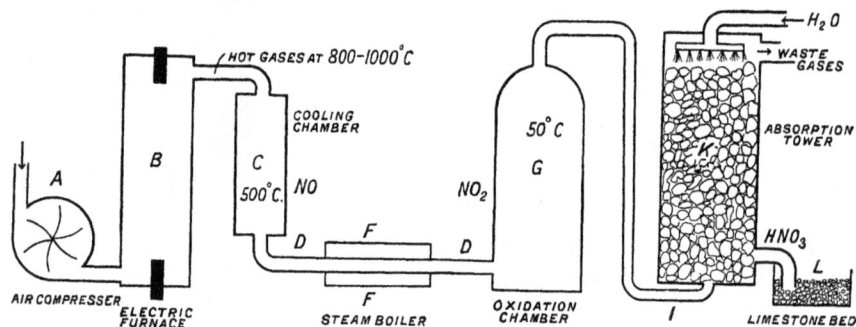

FIG. 9.—Plan of Plant for the Manufacture of Nitrates by the Electrical Oxidation of the Air.

that in practice most of the NO formed is decomposed again), and the gases escape from the furnace at a temperature of 800°-1,000° C. containing only 1.5-2 per cent. by volume of NO, the rest being unchanged oxygen and nitrogen (*e.g.*, 30 mg. HNO₃ per litre). The hot gases at 800-1,000° C. stream into a cooling chamber C, where the temperature falls to about 500° C., and then are passed, still very hot, through a series of tubes contained in a number of tubular boilers. Here they give up their heat to the water in the boilers, the latter developing a pressure of 130 lbs. on the square inch, and thus sufficient steam is generated to work all the necessary machinery in the works, the use of coal being thus entirely done away with. The gases leave the boilers at a temperature of only 200°-250° C., and by passing through another cooling arrangement their temperature is further lowered to about 50° C., when they finally enter a large oxidation chamber G, consisting of a series of vertical iron cylinders with acid-proof lining. Now the nitric oxide only begins to unite ($2NO + 2O_2 = 2NO_2$) with the oxygen of the excess air at a temperature below 500° C., and so the formation of nitrogen peroxide, NO₂, already begins in the gases, and this oxidation is nearly completed in the large oxidation chamber G, the gases finally leaving G consisting of 98 per cent. of air, and 2 per cent. of a mixture of 75 per cent. NO₂ and 25 per cent. NO.

Next the gases pass from the oxidation chambers into a series of absorption towers K (one tower only being shown), where they meet a stream of descending water, and the nitrous peroxide is absorbed with the formation of nitric and nitrous acids, thus :—

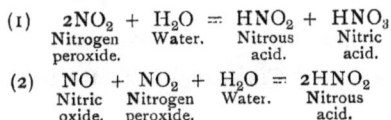

(1) $\underset{\substack{\text{Nitrogen} \\ \text{peroxide.}}}{2NO_2} + \underset{\text{Water.}}{H_2O} = \underset{\substack{\text{Nitrous} \\ \text{acid.}}}{HNO_2} + \underset{\substack{\text{Nitric} \\ \text{acid.}}}{HNO_3}$

(2) $\underset{\substack{\text{Nitric} \\ \text{oxide.}}}{NO} + \underset{\substack{\text{Nitrogen} \\ \text{peroxide.}}}{NO_2} + \underset{\text{Water.}}{H_2O} = \underset{\substack{\text{Nitrous} \\ \text{acid.}}}{2HNO_2}$

According to these equations much nitrous acid is formed. However, in practice, *with increasing concentration, the nitrous acid becomes unstable* and decomposes into nitric acid and nitric oxide, thus :—

$$3HNO_2 = HNO_3 + 2NO + H_2O.$$

Under certain conditions of temperature, and by diminishing the amount of water, the instability of nitrous acid is so marked that practically only nitric acid is finally left in solution.

The absorption of the nitrous fumes takes place in such a way that the first absorption tower contains the strongest acid and the last the weakest acid, the practical result being that the first tower not only contains the strongest nitric acid, but also an acid practically free from nitrous acid. The NO passes on from absorption tower to absorption tower, and is oxidised to NO_2 and absorbed according to the equations given above. Hence the amount of nitrous acid, HNO_2, increases in each succeeding tower, until in the fifth tower (which is fed with a solution of sodium carbonate) only pure sodium nitrite is formed.

The absorption towers consist of three series of stone towers, 20 m. high and 6 m. internal diameter. Each series contains five towers, viz., three acid and two alkali towers. The acid towers are built of granite slabs bound together by iron rods, having a capacity of 600 cub. m. They are filled with broken quartz, over which water (or the dilute nitric acid formed) is slowly flowing. The alkali towers are built of wood bound together by wooden rods, and having a capacity of 700 cub. m.; they also are filled with broken quartz, down which flows a solution of sodium carbonate. The passage of the gases from tower to tower are aided by aluminium fans.

The water is allowed to flow down the third absorption tower (being once more pumped to the top when it reaches the bottom) until it attains a strength of 5 per cent. HNO_3 by volume. This liquid is then pumped to the top of the second tower, where it is allowed to circulate until it attains the strength of 20 per cent. HNO_3 by volume. Next the acid is pumped to the top of the first tower, where, meeting the fresh gases from the oxidation chambers, it attains the strength of 40-60 per cent. HNO_3 by volume, and thus, nitrous acid being under these conditions unstable, the first tower is made to contain not only the strongest nitric acid, but also an acid free from nitrous acid (see above).

The tower coming after the third tower is the alkali tower, being fed with sodium carbonate solution, and in this fourth tower a mixture of sodium nitrate and nitrite is produced (sodium nitrate). The fourth tower, likewise fed with sodium carbonate solution, produces practically pure sodium nitrite, as above explained.

The total absorption in these towers is over 98 per cent.

The main product of manufacture is the 40 per cent. by volume nitric acid obtained from the first tower. 1 kilowatt-year gives 550-650 kilos of HNO_3 (calculated as 100 per cent.)

Five finished products are made in the works :—

(1) **40 per cent. HNO_3** from the first tower.

(2) **Calcium nitrate**, $CaNO_3$, made by running the crude nitric acid through a series of granite beds filled with limestone ($CaCO_3$) until the liquid contains under 0.5 per cent free HNO_3. The liquor is neutralised with lime, evaporated in vacuum pans until of 1.9 sp. gr., and allowed to solidify. It is then either exported in drums, or ground up and put into casks, being sold as "Norwegian saltpetre." It is reddish-brown to black. It is further described on p. 12.

(3) **Sodium Nitrite**, $NaNO_2$.—The liquid from the fifth absorption tower is evaporated, run into shallow tanks, allowed to crystallise, centrifuged, and the crystals dried in a current of hot air. The product contains 99 per cent. $NaNO_2$, being sold as a fine white powder (see p. 14 for properties).

Nitrites are also made from the *furnace gases* by passing the latter, still at a temperature of 200°-300° C. directly up an absorption tower fed with sodium carbonate solution. Under these conditions only nitrite is produced according to the equation :—

$$Na_2CO_3 + NO + NO_2 = 2NaNO_2 + CO_2.$$

In the experimental factory at Christiansand all the nitrite produced is made by this process.

(4) **Sodium Nitrite Nitrate.**—A yellow crystalline material produced by the first alkali tower. It contains 50 per cent. $NaNO_2$, 43 per cent. $NaNO_3$, and 7 per cent. H_2O, together with a little unchanged sodium carbonate. The substance is used instead of Chile saltpetre for the manufacture of sulphuric acid by the chamber process.

(5) **Ammonium Nitrate**, NH_4NO_3.—Produced by neutralising the 40 per cent. HNO_3 directly with ammonia liquor of 0.880 sp. gr. The liquid is evaporated to 1.35 sp. gr., is allowed to crystallise, centrifuged, dried in hot air, when they contain 99.9 per cent. NH_4NO_3.

Manufacture of Nitric Acid from Synthetic Ammonia

(The Ostwald Process.)

Recently ammonia, NH_3, has been manufactured extremely cheap by several methods, *e.g.*, directly from its elements, also by passing steam over nitrolime. It has, consequently, greatly fallen in price, and has thus become available as the starting-point for the manufacture of nitric acid.

Kühlmann in 1830 oxidised ammonia, using platinum as a catalyst, but ammonia was in those days too expensive for practically producing nitric acid by this process. About 1900 Prof. Wilhelm Ostwald, together with his assistant, Dr Brauer (see *Chem. Zeitung.*, 1903, 457, also English Patent, 698, 1902 ; 8,300, 1902 ; 7,909, 1908 ; American Patent, 858,904, 1907 ; Ostwald, *Berg. u. Huttenm. Rundschau*, 1906, **3**, 71 ; Schmidt u. Böcker, *Ber.*, 1906, p, 1366), re-investigated the process, and an experimental plant was erected in which some 150 tons of dilute nitric acid were produced per month. Even at that time, however, ammonia seems to have been too expensive for the process to compete with the sodium nitrate process of manufacture.

Recently, however, ammonia has been produced at one-fourth its former price (by the synthetic process), and so the difficulty as regards the first cost of ammonia appears to have vanished.

Also the process in its original form was imperfect. The technical details were gradually improved, and in 1909 a factory was built at Gerte, Westphalia. Numerous other factories are now being erected. Works exist in Belgium, near Dagenham on the Thames, Trafford Park, Manchester, in Scotland and Ireland. Rudolph Messel (*Journ. Soc. Chem. Ind.*, 1912, **31**, 854) has pointed out that the production of nitric acid from the air (see above) is at present confined to countries producing water power, and that the resulting nitric acid is very difficult to transport ; on the other hand, ammonia is producible almost everywhere (since not much energy is required in its production), and can, with an Ostwald plant, be readily oxidised on the spot to nitric acid. Consequently this new process, especially since the production of synthetic ammonia, may ultimately prove the most important technical method for producing nitric acid.

For further details see Donath and Indra, " Die Oxydation des Ammoniaks zu Salpetersäure und Salpetriger Säure" (Stuttgart, 1913) ; see also under **Ammonium Nitrate**, p. 13.

Under the influence of a catalyst, ammonia, NH_3, can be oxidised by air to form water and nitric acid or oxides of nitrogen ; thus :—

$$(1)\ NH_3 + 2O_2 = HNO_3 + H_2O.$$

However, unless certain conditions are maintained, the oxidation may be incomplete and only nitrogen gas be produced, thus :—

$$(2)\ 4NH_3 + 3O_2 = 2N_2 + 6H_2O.$$

This formation of nitrogen was one of the chief difficulties to be overcome.

Now, in order to obtain a technically useful result, the operation must be so conducted that the first reaction (1) is practically complete, whilst (2) must be as small as possible. This result is attained by using, for example, smooth or solid platinum, which causes the first action, viz., the production of nitric acid, to take place almost quantitatively, the production of free nitrogen being unnoticeably small. However, the action is slow with the use of smooth platinum alone. On the other hand, finely divided platinum, or platinum black, accelerates *both* reactions, but accelerates the second one (*i.e.*, the production of free nitrogen) more than the first. However, by moderate use of the finely divided platinum (platinum black) with the smooth platinum the operation can be so performed that the action takes place rapidly, but without any great formation of free nitrogen.

W. Ostwald, therefore, in his English Patent, 698, 1902, advised that the platinum should be partly covered with platinum black, and that the gas current should pass at the rate of 1-5 m. per second through a length of 1 cm. or 2 cm. of contact substance maintained at about 300° C.

It is unnecessary to heat the platinum, since heat is liberated in the course of the reaction, which raises the temperature of the platinum to a sufficient extent.

The actual contact between the gas and the platinum should not last longer than one hundredth second, and so a rapid stream of gas is essential in order to fulfil this condition.

The ammonia is mixed with about 10 vols. of air in order to conform to the equation :—

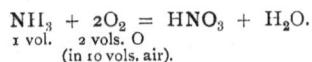

$$NH_3 + 2O_2 = HNO_3 + H_2O.$$
$$\underset{\text{1 vol.}}{} \quad \underset{\substack{\text{2 vols. O} \\ \text{(in 10 vols. air).}}}{}$$

The catalysts are arranged in a kind of semi-parallel system, so as to allow the cutting out of any section. The nitric acid vapours are condensed in towers in which they meet nitric acid ; the addition of water is unnecessary, since it is formed during the oxidation of the ammonia.

The plant in operation at the Lothringen Colliery Co., near Bochum, were in 1912 producing 1,800 tons of nitric acid and 1,200 tons of ammonium nitrate annually. At these works the carefully purified ammonia mixed with 10 vols. of air is sent through enamelled iron pipes into a chamber containing the special platinum catalyst. The nitric acid vapours here produced pass through aluminium tubes to an absorption plant, consisting of six towers packed with broken earthenware, down which a stream of nitric acid trickles, being continually pumped again to the top of the tower as soon as it reaches the bottom. Here acid forms which produces directly a nitric acid containing 55 per cent. HNO_3. By altering the condensing arrangements a nitric acid of 66 per cent., or even 92 per cent., can be produced as a sole product, and of a purity sufficient for its direct utilisation for the manufacture of explosives. About 85-90 per cent. of the ammonia is thus transformed into nitric acid.

In the works above mentioned the nitric acid is neutralised by ammonia, and thus ammonium nitrate is produced in bulk.

Other catalysts besides platinum will oxidise ammonia. Thus Frank and Caro (German Patent, 224,329) showed that the expensive platinum can be replaced by a mixture of ceria and thoria, which at 150°-200° C. give a yield of 90 per cent. HNO_3 or HNO_2. The mixture is not so efficient as platinum, but is far cheaper.

Fr. Bayer & Co. (German Patent, 168,272) showed that at 600°-750° burnt pyrites (iron oxide containing some copper oxide (see p. 13), oxidises ammonia, N_2O_3, being the sole product of the oxidation. By absorbing in alkali, this gave nitrite ; see also under **Ammonium Nitrate**, p. 13.

The ammonia used may be produced from mixed gas acid and gas liquor, since, although some purification with lime is necessary, the organic impurities in it do not materially affect the course of the reaction.

However, much cheaper sources of ammonia gas are calcium cyanamide (which yields NH_3 and $CaCO_3$ when treated with superheated steam, $Ca=N-RN+3H_2O = CaCO_3 + 2NH_3$), and synthetic ammonia produced by the Badische Anilin- und Soda Fabrik by direct union of N and H (see p. 53).

The Nitrates Products Co. Ltd. have secured the world's rights of this process outside Westphalia and Rhineland. They intend to produce their ammonia from the action of steam on nitrolime (calcium cyanamide).

The Badische Anilin- und Soda Fabrik will probably produce nitric acid, using their synthetic ammonia (see below).

Properties of Nitric Acid.—The strong nitric acid, possessing a specific gravity of 1.50-1.52, is a powerful oxidising agent. In fact, it will spontaneously inflame wood or straw if poured upon them in quantity, and, consequently, is always technically transported in glass balloons, with a packing of sand or kieselguhr. It is usually only sent by special trains, much in the same way that explosives are sent.

Owing to the difficulty of transport, concentrated HNO_3 is usually made on the place of consumption, *e.g.*, it is far more economical to manufacture nitric acid directly in the neighbourhood where it is to be consumed than to manufacture it in distant countries, where the conditions are more favourable, and then send it long distances by sea or rail to the point of use—as is nearly always done in the case of other heavy chemicals, such as sodium sulphate or alkali.

Concentrated nitric acid is used, mixed with concentrated sulphuric acid, for nitrating purposes, especially for the manufacture of nitroglycerine and nitrocellulose, collodion and celluloid.

For example, for the manufacture of nitroglycerine a mixture of 36.5 per cent. HNO_3 (of 91.6 per cent. strength) and 63.5 per cent. H_2SO_4 (of 97 per cent. strength), together with 5 per cent. H_2O is used. Such an acid must be free from chlorine, and must not contain more than 0.2-0.3 per cent. of nitrous acid.

For making such substances as nitrobenzene, picric acid, and other organic nitrogenous bodies, a weaker acid of 1.35-1.40 sp. gr. mixed with 1-2 parts of sulphuric acid (66° Bé., 67° Tw.) is used.

A nitric acid of 1.35-1.40 sp. gr. is also used for the Glover towers in the manufacture of sulphuric acid.

A more dilute nitric acid is used for making silver nitrate and other nitrates, for etching copper plates, and for dissolving metals.

The *red fuming nitric acid*, of specific gravity 1.4-1.5, contains dissolved in it oxides of nitrogen, and is a more powerful oxidising medium than the colourless acid of the same specific gravity.

This acid is made sometimes by adding a little starch to strong nitric acid, which causes its partial reduction, and sometimes by distilling sodium nitrate with a smaller amount of sulphuric acid, and at a higher temperature than is required for the production of the ordinary acid.

The following table, after Lunge and Rey (*Zeit. angew. Chem.*, 1891, 167 ; 1892, 10), gives the strengths of solutions of nitric acid :—

Specific Gravity.	Baumé.°	100 g. Contain g. HNO_3.	100 c.c. Contain g. HNO_3.
1.000	0	0.10	0.10
1.025	3.4	4.60	0.47
1.050	6.7	8.99	9.7
1.075	10.0	13 15	14.1
1.100	13.0	17.11	18.8
1.125	16.0	21.00	23.6
1.150	18.8	24.84	28.6
1.175	21.4	28.63	33.6
1.200	24.0	32.36	38.8
1.225	26.4	36.03	44.1
1.250	28.8	39.82	49.8
1.275	31.1	43.64	55.6
1 300	33.3	47.49	61.7
1.325	35.5	51.53	68.3
1.350	37.4	55.79	75.3
1.375	39.4	60 30	82.9
1.400	41.2	65.30	91.4
1.425	43.1	70.98	101.1
1.450	44.8	77.28	112.1
1.475	46.4	84.45	124.6
1.500	48.1	94.09	141.1
1.510	48.7	9).10	148.1
1.520	49.4	99.67	151.5

Statistics.—According to the *Report on First Census of Production*, 1907, p. 571, the United Kingdom in 1907 produced 60,000 tons of nitric acid, valued at £91,000. The exports are shown by the following figures :—

1908.		1909.		1910.	
Tons.	Value.	Tons.	Value.	Tons.	Value.
257	£5,468	269	£7,093	185	£4,972

Germany produces annually about 100,000 tons of HNO_3 calculated as 100 per cent.

CHAPTER IV

The Ammonia and Ammonium Salts Industry

CHAPTER IV

THE AMMONIA AND AMMONIUM SALTS INDUSTRY

For Synthetic Ammonia see Chapter III.

LITERATURE

LUNGE.—"Coal Tar and Ammonia."

R. ARNOLD.—"Ammonia and Ammonium Compounds."

G. T. CALVERT.—"The Manufacture of Sulphate of Ammonia." 1911.

CARO.—"Die Stickstoffrage in Deutschland." Berlin, 1908. *Chem. Zeit.*, 1911, 505.

FRANK and CARO.—*Zeit. angew. Chem.*, 1906, 1569.

WOLTERECK.—English Patents, 16,504, 1904; 28,963, 1906; 28,964, 1906.

BEILBY.—*Journ. Soc. Chem. Ind.*, 1884, 216.

NORTON.—"Consular Report on the Utilisation of Atmospheric Nitrogen." Washington, 1912.

DONATH and INDRA.—"Die Oxydation des Ammoniaks zu Salpetersäure und Salpetriger Säure." 1913.

Also references and patents given in text; *see also* under *Synthetic Ammonia*, p.   .

Ammonia and Ammonium Salts.—Ammonia, NH_3, is a decomposition product of organic matter, resulting either from its destructive distillation or from its putrefaction and decay.

Until quite recently, practically the total supply of ammonia was obtained as a by-product in the destructive distillation of coal for the purpose of making illuminating gas, much also being produced by coke ovens, and the processes now worked for the gasification of coal, peat, oil-shale, etc.

The amount of nitrogen in coal amounts to 1·2 per cent., and only a relatively small proportion of this (from 12-20 per cent. of the total nitrogen present) is converted into ammonia during the process of destructive distillation or gasification. At least half of the nitrogen remains in the residual coke, and is not expelled completely even at a white heat. Much nitrogen escapes from the coal in the form of N_2 gas, being formed by the decomposition of the NH_3, which begins to decompose at 600°-800° C. A small proportion of the nitrogen distils over in the form of cyanide, pyridine, and other organic nitrogenous substances. The following figures show how the nitrogen contained in various sorts of coal is disposed of during the ordinary process of destructive distillation of 100 parts of their total N. We have:—

Kind of Coal.	English.	Westphalian.	Saar.
	Per Cent.	Per Cent.	Per Cent.
N in coke - - -	48-65	30-36	64
N as NH_3 - - -	11-17	11-15	16
N as gaseous N_2 - -	21-35	47·55	16
N as cyanide - - -	0.2-1.5	1-2	} 4.0
N in tar - - -	1-2	1-1.5	

In general, 100 kilos of average coal on destructive distillation yield 0.25-0.3 kg. of ammonia or 1.0-1.2 kg. of ammonium sulphate. When, however, the coal undergoes destructive distillation *in a stream of superheated steam* (as in the processes of producing "water gas" or "Mond gas" from crude coal) we can obtain as much as 3 kilos of ammonium sulphate per 100 kilos of coal treated.

The following may be considered as the chief sources of ammonia and ammonium salts, as they at present exist :

(1) **The Coking of Coal.**

(*a*) **For the Production of Metallurgical Coke.**—Already in Germany no less than 90 per cent. of total output of ammonium sulphate is manufactured in recovery ovens ; in England some 60 per cent., and in the United States only 20 per cent. is thus obtained. About 55 per cent. of the world's total output of ammonia in 1911, and 74 per cent. of that of the U.S.A., is derived from the carbonisation of coal in by-product ovens. There is little doubt that the output will increase as the amount of coal carbonised increases.

(*b*) **For the Production of Coal-Gas.**—30 per cent. of the world's output in 1910 was obtained from the retorting of coal for the manufacture of illuminating gas.

(2) **The Distillation of Shale.**—Large amounts are recovered in Scotland by the distillation of bituminous shale. The following figures, taken from the "49th Annular Report on Alkali, etc., Works," p. 130, show this :—

Year.	Total Shale Mined and Quarried in Scotland.	Total Sulphate of Ammonia Recovered from the Shale in Paraffin Oil Works.	Yield in Lbs. per Ton of Shale.
	Tons.	Tons.	
1903	2,009,265	37,353	41.6
1904	2,331,885	42,486	40.8
1905	2,493,081	46,344	41.6
1906	2,545,724	48,534	42.7
1907	2,690,028	51,338	42.7
1908	2,892,039	53,628	41.5
1909	2,967,017	57,048	43.1
1910	3,130,280	59,113	42.3
1911	3,116,803	60,765	43.7
1912	3,184,826	62,207	43.7

Considerable undeveloped deposits of bituminous shale exist in Newfoundland, Australia, and other parts of the world.

(3) **The Distillation or other Treatment of Peat.**—Much nitrogen is combined in peat (which, therefore, has found some application as a fertiliser), and many processes are either worked or are projected for directly transforming this nitrogen into the form of ammonium sulphate. The destructive distillation of peat has been proposed by Ziegler and others.

More promising is the partial combustion of peat with production of producer gas (Frank, Caro, Mond, described on p. 43) or the slow wet-combustion of peat (Woltereck, p. 43). There are 20,000 million tons of undeveloped peat in the U.S.A., while equally enormous quantities exist in Ireland, Canada, Newfoundland, Sweden, Norway, Russia, Germany (Prussia alone containing 5,000,000 acres).

The average nitrogen content of these peat deposits may be taken as 2.05 per cent. (sometimes reaching 4 per cent. in the case of dry peat), so that if only 50 per cent. of this was recovered, very large amounts of ammonium sulphate would become available.

(4) **Producer Gas.**—The ammonia contained in the coals used in producing this gas—the main type of plant used being the Mond Gas Producer—is now very large and is likely to increase. The process is described in Martin's "Industrial Chemistry," Vol. II.

(5) **Blast-Furnace Gas.**—The nitrogen in the coal used in blast furnaces escapes in part as ammonia, much, however, being decomposed by the high temperatures in the furnace.

Some of the combined nitrogen, however, comes from the nitrogen of the air by actions taking place inside the furnace, principally the formation of cyanides, which are then decomposed into ammonia by aqueous vapour in the furnace. In the United Kingdom, for 1910, about 20,130 tons of ammonium sulphate were recovered, 20,130 tons being recovered in 1911, and 17,026 in 1912.

(6) **Production from Beetroot Sugar Waste, "Schlempe" or "Vinasse."**—The thick brown liquid remaining after the extraction of all the possible sugar from the syrup is know as "vinasses" or "schlempe."

It contains much nitrogen and potassium salts. Until recently it was the custom merely to calcine this material so as to obtain the potash salts in the form of "schlempe kohle." Bueb of Dessau and Vennator now recover nitrogen from this by distilling the schlempe from iron retorts, leading the evolved vapours through chambers filled with brick cheque-work maintained at a red heat, whereby the complex vapours decompose into HCN, NH_3, etc.

The ammonia and cyanogen are then recovered as in the purification of coal gas. The process is described on p. 76, "Cyanide and Prussiate Industry."

(7) By Distillation of Bones, Leather, and other Nitrogenous Organim Matter.—The distillation of bones, for the production of "bone black," formerly yielded a considerable supply of ammonia, but the industry is now not so prominent as formerly.

The ammonia may be extracted from the evolved gases by scrubbing, as in coal gas manufacture. The treatment of animal refuse for the manufacture of prussiate is now obsolete. It is more profitably employed as manure. See p. 79, "Cyanide and Prussiate Industry."

(8) **From Sewage and Urine.**—A very rich source of ammonia is ordinary urine. 100,000 heads of population could produce per year about 6,000 tons of NH_3. If all the ammonia corresponding to London urine were collected, more than 60,000 tons of ammonium sulphate could be annually produced therefrom.

The method of collection of urine and its working up into ammoniacal compounds has been carried on at Paris and at Nancy. In 1909 France obtained 13,000 tons of ammonium sulphate therefrom, 10,000 being obtained in Paris alone. However, the collection and utilisation of animal excrement is so nauseous and costly and dangerous a process, that the bulk of the enormous ammonium supplies producible from this source are run to waste.

The process consists in allowing the urine to ferment into ammonium carbonate. The clear liquor is distilled and the ammonia recovered as in gas-liquor.

For further details see Ketjen, *Zeit. angew. Chem.*, 1891, 294; Butterfield and Watson, English Patent, 19,502/05; Taylor and Walker, U.S. Patent, 603,668; Young, English Patent, 3,562/82; Duncan, German Patents, 27,148, 28,436.

(9) **Synthetic Ammonia.**—Enormous supplies of ammonia are now becoming available by the synthesis of ammonia, either directly from atmospheric N and H, or else from cyanamide or nitrides. These processes are discussed in detail in a separate article, p. 53.

The production of ammonium sulphate is increasing rapidly in order to meet the increasing demand for nitrogenous manures. The following figures refer to Great Britain :—

AMOUNT OF AMMONIA RECOVERED IN THE UNITED KINGDOM (EXPRESSED IN TERMS OF SULPHATE)—TONS.

	1910	1911.	1912.	1913.	1914.
Gas works	167,820	168,783	172,094	182,000	177,000
Iron works	20,139	20,121	17,026	20,000	19,000
Shale works	59,113	60,765	62,207	63,000	62,000
Coke oven works	92,665	105,343	104,932	...	...
Producer-gas and carbonising works (bone and coal)	27,850	29,964	32,049	167,000	163,000
Total	367,587	384,976	388,308	432,000	421,000

The following figures refer to the output of the chief countries :—

	1900.	1909.	1911.
	(Metric) Tons.	(Metric) Tons.	(Metric) Tons.
England	217,000	349,c00	378,000
Germany	104,000	323,000	400,000
United States	58,000	98,000	127,000
France	37,000	54,000	60,000
Belgium, Holland Austria, Russia, etc.	} 68,000	134,000	...

The **world's production** of ammonium sulphate is estimated as :—

1900.	1909.	1911.
484,000 tons.	950,000 tons.	1,150,000 tons.

Great Britain exported ammonium sulphate :—

1911.	1912.	1914.
292,000 tons.	287,c00 tons.	314,000 tons.

The chief product of the ammonia industry is, at present, *solid ammonium sulphate*, $(NH_4)_2SO_4$, which is principally used in agriculture as a manure. It is valued on the percentage of nitrogen it contains, containing when pure, as $(NH_4)_2SO_4$, about 21.2 per cent. N against 16.5 per cent. N in $NaNO_3$, or Chile saltpetre, which is at present its great competitor as a manure.

The price of ammonium sulphate sunk from 50s. per ton in 1880 to 20s. to 30s. in 1909, and no doubt in consequence of the production of cheap synthetic ammonia, both from cyanamide, and by direct synthesis, the price will probably still further decrease.

Manufacture of Ammonium Sulphate from Gas - Water or Ammoniacal Liquor.—At present the bulk of the ammonium sulphate on the market is derived from the "gas-water" or "*ammoniacal liquor*" produced in the numerous coal-gas producing plants, coke-ovens, etc.

In the coke-ovens, however, the gases now are passed directly through sulphuric acid, and the ammonium sulphate thereby directly fixed, thus avoiding the initial production of an "*ammoniacal liquor*" such as is indispensable to coal-gas production.

Ordinary ammoniacal liquors contain some 1.5-3 per cent. NH_3, united with various acids. The chief acid is carbonic, H_2CO_3, but besides this we get H_2S, HCN, $HCNSH_2S_2O_3$, H_2SO_4, HCl, and ferro- and ferri-cyanic acids.

The ammonium salts are, in practice, divided into (*a*) volatile, (*b*) fixed. The "volatile" ammonium salts on boiling with water dissociate, evolving ammonia. The chief volatile salts are :—Ammonium carbonate $(NH_4)_2CO_3$; ammonium sulphide $(NH_4)_2S$ and NH_4HS ; ammonium cyanide, NH_4CN.

The "fixed" ammonium salts (*e.g.*, ammonium sulphate $(NH_4)_2SO_4$; ammonium chloride (NH_4Cl, etc.), are not decomposed by water, but the ammonia has to be driven out of them by boiling with milk of lime.

Different ammoniacal liquors, however, have an extremely variable composition. An average sample would contain per 100 c.c. from 1.4-3.3 g. volatile ammonia (principally in the form of ammonium carbonate), and 0.2-0.6 g. NH_3 in the form of "fixed" salts. The composition of the ammoniacal liquor naturally largely depends upon the nature of the coal used, and some coals, rich in chlorides, yield ammoniacal liquors containing much NH_4Cl.

The following analyses gives the composition of some average ammoniacal liquors, the numbers giving grams per 100 c.c. :—

	Gas Works.	Coke Ovens.	Blast Furnaces.	Shale Works.	Coalite Works.
Volatile NH_3 - - -	1.4-3.3	0.84	0.2-0.4	0.9	1.5
Fixed NH_3 - - -	0.6-0.2	0.10	0.008-0.009	0.03	0.17
Total NH_3 - - -	2.5-3.5	0.94	0.2-0.4	0.9	0.7
$(NH_4)_2S$ - - -	0.9-0.8	0.47	...	0.1	0.23
$(NH_4)_2CO_3$ - - -	5.0-8.8	1.96	1.1	2.9	6.4
$(NH_4)Cl$ - - -	1.1-0.5	0.22	0.006	0.015	0.1
$(NH_4)_2SO_4$ - - -	0.2-0.0	0.03	0.009	0.016	0.05
$(NH_4)_2S_2O_3$ - - -	0.17-0.0	0.04	0.002	0.09	0.4
NH_4CNS - - -	0.53-0.07	0.04	0.003	...	0.3
NH_4CN - - -	0.036-0.07	0.07	0.003	...	...
$(NH_4)_4Fe(CN)_6$ - - -	0.038	...	...	...	...

Organic substances such as phenol, pyridin, acetonitrol, etc., also occur in small amounts.

The method of working the ammoniacal liquor for ammonium sulphate is first to boil it until all the "volatile" ammonium salts have been distilled off. To the residual liquid containing the "non-volatile" ammonium salts the theoretical amount of milk of lime is added, and the boiling continued until all their ammonia is also expelled.

The evolved vapours are usually led directly into H_2SO_4 of 42°-46° Bé. (81°-93° Tw.), and the ammonia fixed in the form of solid ammonium sulphate $(NH_4)_2SO_4$, which can be sold without further refining for manurial purposes.

The plant used consists of "column" apparatus similar to those described in detail in Vol. I. "Organic Industrial Chemistry," for distilling alcohol; the apparatus, however, is modified so that the ammoniacal liquor alone is distilled in the upper part of the apparatus so as to expel all volatile ammonia; while in the lower part of the apparatus the "fixed" ammonium salts in the residual liquid are boiled with milk of lime.

There are a great many different plants on the market, some of which are extremely efficient. **Feldmann's Apparatus** (D.R.P., 21,708, see English Patent, 3,643, 1882), shown in Fig. 10.

The ammoniacal "gas-water" flows into a tube from the regulating tank A

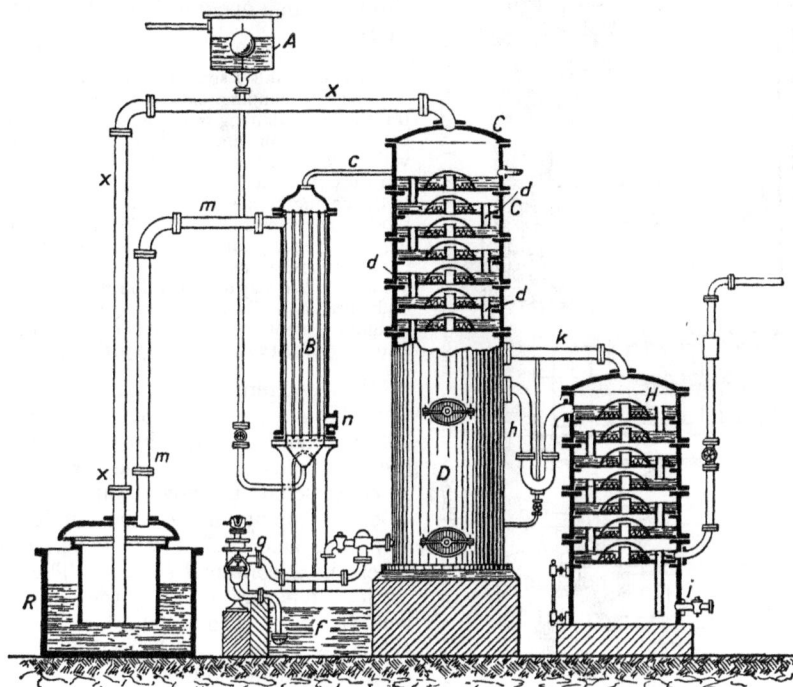

FIG. 10.—Feldmann's Ammonia Still.

and enters the multitubular "preheater" B, consisting of a series of tubes through which the ammoniacal liquor flows, which are themselves heated by the steam and hot gas coming from the saturator R by the pipe MM. From the "preheater" B the now hot ammoniacal fluid flows into the top chamber of the column C. This is provided with a number of compartments each provided with an overflow pipe D, so that in each compartment the liquor accumulates to an appreciable depth. In the centre of the floor of each compartment is a wider pipe covered over with a "bell" or "mushroom" (e), provided with serrated edges (see Martin's "Industrial Chemistry," Vol. II., under "Ammonium Soda Industry"). Through this central pipe the ammoniacal gases and steam come up from below and stream through the liquor surrounding the "mushroom," and thus boil out all the volatile NH_3.

The liquid in C, from which all volatile ammonia has been boiled out, now enters the lower part of the still D. Into this compartment a stream of milk of lime is continually pumped by means of the pump g, the lime being sucked out

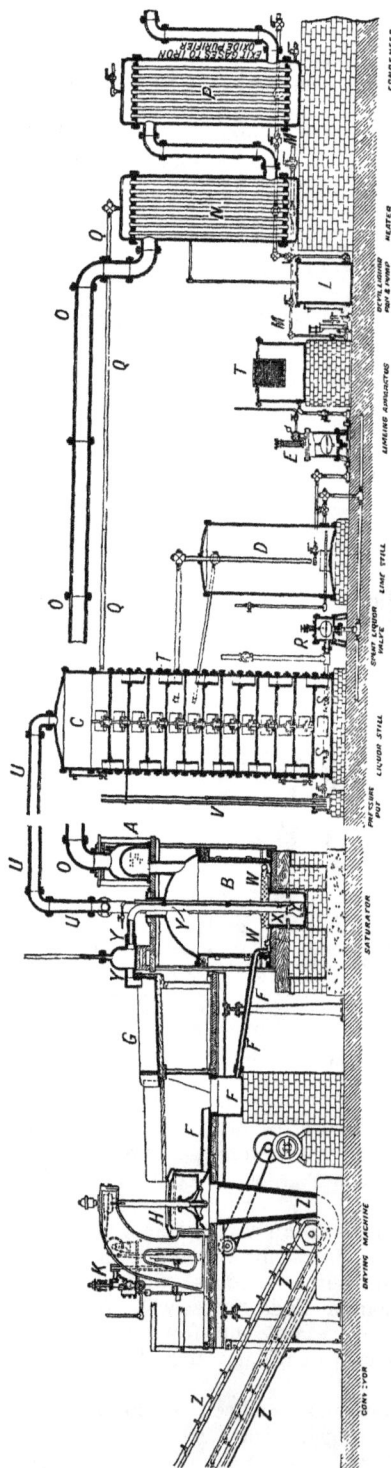

FIG. 11.—Ammonium Sulphate Plant of the Chemical Engineering Co., Hendon, England.

of the tank *f.* The mixed fluids flow through a filtering sieve (to retain large particles) through the tube *h* into the column H, through which is passed from below a current of steam from a boiler. This steam maintains the whole liquid at a boiling temperature and completely expels all the ammonia from the "fixed" ammonium salts, the ammonia being, in the first place, set free by the milk of lime.

The waste ammonia-free liquors run away through *i.*

The mud-like mass of lime escapes through an opening at the bottom of D.

In some plants a special separate mixing vessel is provided, standing outside the column. Into this the liquid coming from the middle part of the column C is run, and is then intimately mixed with milk of lime, and then the mixed fluids are run back into the lower column B, and subjected to the boiling by means of steam. A still of this type is manufactured by the Chemical Engineering Co., Hendon, and is shown in Fig. 11.

Many special modifications of this apparatus are used. We may here mention the apparatus of Grüneberg and Blum (D.R.P., 33,320); Wilton (English Patent, 24,832, 1901); Scott (English Patent, 3,987, 1900; 11,082, 1901).

The escaping steam, carrying with it the NH_3 gas, passes out at the top of the column C and through the tube x x into the lead-lined (or volvic stone) "*saturator*" R constructed as shown, with a leaden "bell" dipping under the surface of sulphuric acid (90° Tw, 45° Bé.) which enters in a continual stream. The NH_3 as it enters unites with the sulphuric acid to form solid ammonium sulphate $(2NH_3 + H_2SO_4 = (NH_4)_2SO_4)$ which separates out in the liquid.

It should be noted also that much heat is evolved by the union of the sulphuric acid and the ammonia in the saturator, the heat of interaction being sufficient not only to maintain the saturator at the boiling point, and compensate for unavoidable losses by radiation, etc., but also to more than evaporate the whole of the water contained in the acid, amounting to 20-30 per cent., so that the apparatus can be washed out from time to time, as necessary, without wasting the washings, or evaporating them down externally.

The saturator just described belongs to the "*partly open*" type, the hot waste gases being led off by the pipe M to the heater B, and the sulphate accumulating on the floor of the saturator, the workman removing the latter as it separates by means of a perforated ladle inserted through the open part of the tank. The crystals of sulphate are then placed on a lead-lined drainer, so that the mother liquors flow back to the saturator.

More often the ammonium sulphate crystals are removed from the saturator by means of a steam discharger (working on the principle of the air lift), which drives it in the form of a coarse mud, together with a considerable amount of mother liquor, on to the drainer, and thence it is passed into the centrifugal machines, the mother liquors invariably running back into the saturator. In this type of plant a closed saturator is used, as described below :—

Fig. 11 shows a modern ammonium sulphate plant, erected by the Chemical Engineering Co., of Hendon, London, N.W., for the Grangetown Gas Works, Cardiff, in 1911. This plant is capable of producing 3 tons of ammonium sulphate per twenty-four hours.

The ammoniacal liquor enters the multitubular heater N at the bottom through H, passes up the pipes' insides, and is thereby heated to the boiling point by the steam and hot waste gases from the saturator B. These hot vapours issue from the saturator at A, pass along the pipe OO, then encircle the tubes in the interior of the heater N, heating the ammoniacal liquor therein contained to the boiling point, and being partially condensed in so doing. The condensed steam (known as **Devil Liquor**, on account of the $H_{2}S$, NH_3, HCN, etc., contained therein) runs off into the sealed closed tank L, and is pumped back through MM, and mixed with the gas liquor, the whole thus passing through the liquor still C, which renders it inodorous, the liquors being finally run to waste after passing through the liquor still in the form of "spent liquor." Any moisture not condensed by the heater N is finally condensed in the condenser P (two are employed) of similar construction to N. The H_2S and CO_2 from the saturator B, however, pass for the most part away from the exit gases, being purified with oxide of iron, and the H_2S recovered is sulphur.

The ammoniacal liquor, heated to boiling in N, passes along the pipe QQ, and enters the liquor still C at the top, and slowly flows by a series of weirs, in a downward direction, through the fifteen chambers to the automatic exit, which is controlled by the spent liquor valve R. V is an arrangement for controlling the pressure in the still. Steam is admitted at the bottom of the still by the perforated pipe SS, and bubbles through the liquor in each chamber, travelling in a reverse direction to the liquor, and carrying with it the ammonia and the gases, H_2S and CO_2, which pass away from the top of the still to the acid saturator B, the H_2S and CO_2 thence escaping along OO to the condensers, and reappear in the "devil liquors" at N, as previously explained, and in the waste exit gases from P.

In order to complete the removal of fixed ammonium, salts from the liquor lime is automatically admitted to the middle chamber of the still at T, D being the auxiliary limeing still (see p. 40), which acts as a reservoir, and retains the lime sludge. The lime is slaked in T with spent liquor from the stills, and forced into the still C at boiling temperature at the required rate, by means of automatic pumps. E is the automatic valve controlling the admission of lime to D. The saturator B is of the round closed type, constructed of 40 lbs. chemical lead. In this saturator the ammonia coming from the still C passes through the pipe UU nearly to the bottom of the saturator, and bubbles through the acid bath by means of the perforated pipe WW. The saturator is continuously fed with sulphuric acid in proportion to the amount of entering ammonia. The chemical action, caused by the ammonia uniting with the acid to form ammonium sulphate, developes sufficient heat to cause vigorous boiling, the uprising steam, together with H_2S and CO_2 from the ammoniacal liquors and some ammonia, escape at a high temperature through the tube OO to the heater N and condensers P, as above explained.

The sulphate of ammonia is deposited in a crystalline form into the well X at the bottom of the saturator B, from which it is pumped by means of the steam discharger YY to the receiving tray G, whence it gravitates at intervals to the centrifugal machine (described in Martin's "Industrial Chemistry," Vol. I.), the tray holding a charge of about 4 cwts. The centrifugal machine separates the mother liquor, which flows back to the saturator by FFF. After two minutes' spinning the centrifugal machine H is stopped, and the dry sulphate of ammonia is dropped through the centre valve on to the elevator conveyor ZZ, and deposited in the store.

The apparatus used for the manufacture of ammonium sulphate by the Coppée Gas Co. is shown in Fig. 12. The diagram, Fig. 13, explains the mode of working of the apparatus.

The crystals of sulphate thus obtained are sometimes washed with a very little water, dried, and put on the market for manurial purposes, containing 25.1-25.3 per cent. NH_3. $(NH_4)_2SO_4$ requires $NH_3 = 25.8$ per cent. less than 0.4 per cent. free H_2SO_4, and no cyanide, since the latter is very injurious to vegetation.

Chemically pure ammonium sulphate is sometimes, although rarely, obtained from this raw ammonium sulphate by crystallisation.

Treatment of the Waste Exit Gases from the Ammonium Sulphate Plant.—The acid gases, such as H_2S, HCN, and CO_2, which pass with the ammonia into the sulphuric acid in the

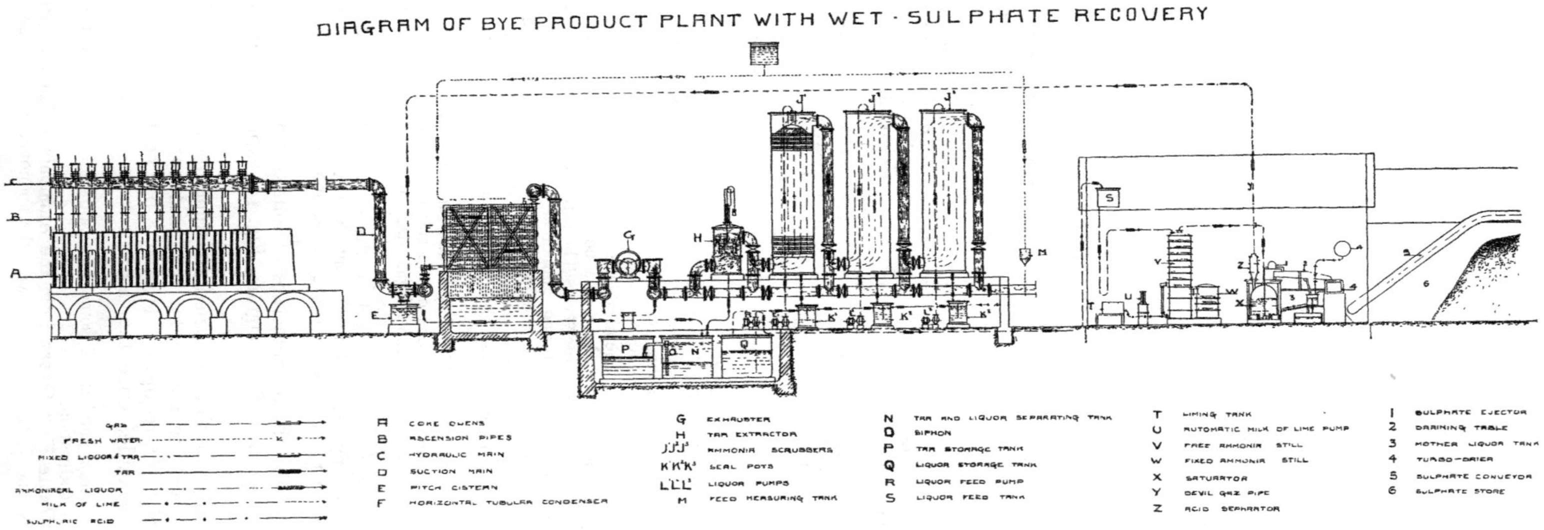

DIAGRAM OF BYE·PRODUCT PLANT WITH WET·SULPHATE RECOVERY

GAS ———————
FRESH WATER ——————
FIXED LIQUOR ————— ·×· ————
MIXED LIQUOR ————
TAR ————
AMMONIACAL LIQUOR ——— ·· ——
MILK OF LIME ——— · — · —
SULPHURIC ACID ——— · ·· — · ·· —

A COKE OVENS
B ASCENSION PIPES
C HYDRAULIC MAIN
D SUCTION MAIN
E PITCH CISTERN
F HORIZONTL TUBULAR CONDENSER

G EXHAUSTER
H TAR EXTRACTOR
J J' J" AMMONIA SCRUBBERS
K K' K" SEAL POTS
L L' L" LIQUOR PUMPS
M FEED MEASURING TANK

N TAR AND LIQUOR SEPARATING TANK
O SYPHON
P TAR STORAGE TANK
Q LIQUOR STORAGE TANK
R LIQUOR FEED PUMP
S LIQUOR FEED TANK

T LIMING TANK
U AUTOMATIC MILK OF LIME PUMP
V FREE AMMONIA STILL
W FIXED AMMONIA STILL
X SATURATOR
Y DEVIL GAS PIPE
Z ACID SEPARATOR

1 SULPHATE EJECTOR
2 DRAINING TABLE
3 MOTHER LIQUOR TANK
4 TURBO·DRIER
5 SULPHATE CONVEYOR
6 SULPHATE STORE

THE COPPÉE COMPANY (GREAT BRITAIN) LTD., KING'S HOUSE, KINGSWAY, W.C.

FIG. 12.—Bye-Product Plant with Wet Sulphate Recovery.

42

Fig. B
Gas Scrubbing

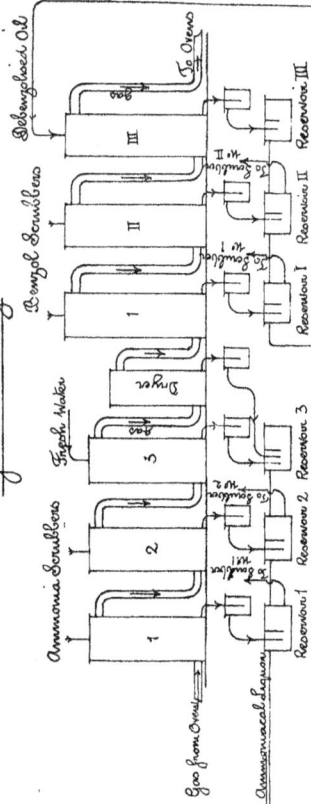

Fig. A
Sulphate Plant (Ordinary method of recovery)

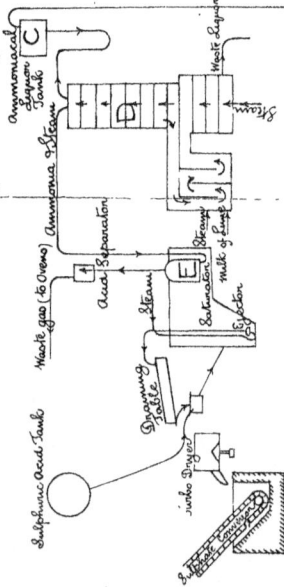

Fig. C
Crude Benzol Plant

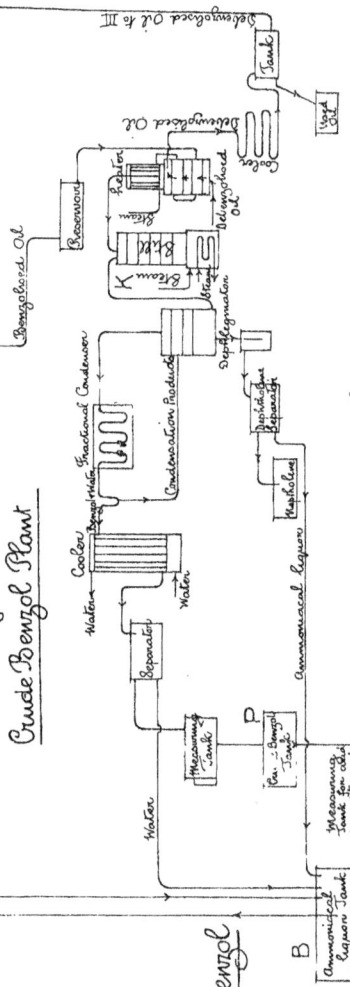

Fig. D
Benzol Rectifying Plant

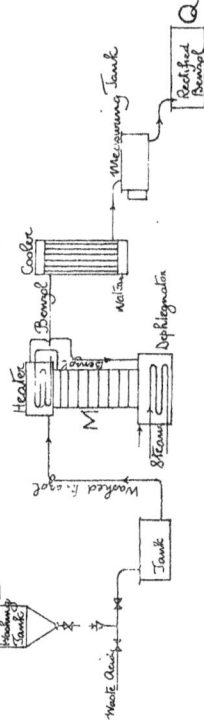

Diagrammatic Arrangement of
Complete Plant for recovery of Ammonia & Benzol

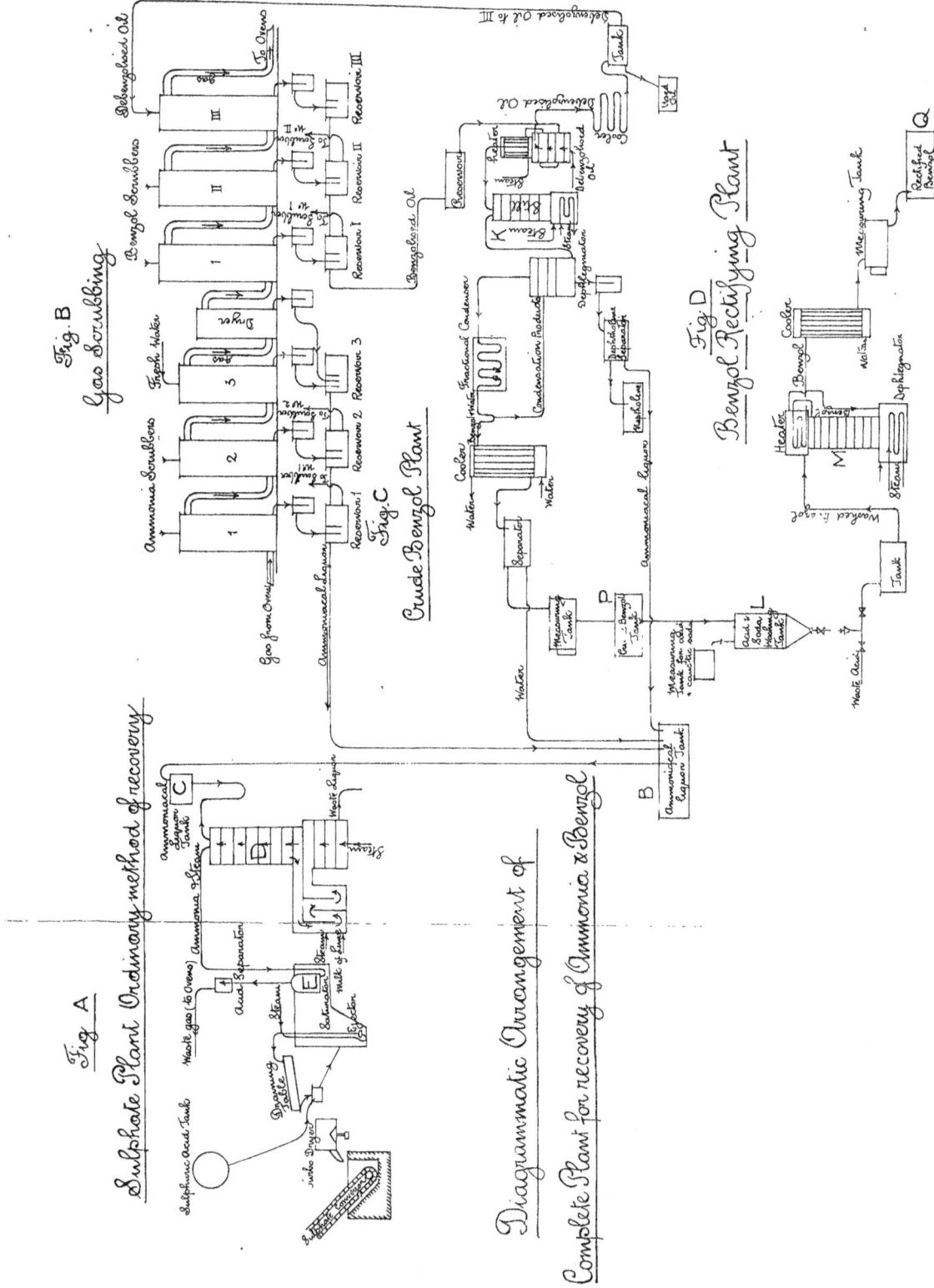

FIG. 13.—PLANT FOR RECOVERY OF AMMONIA AND BENZOL.
(By Permission of the Coppée Co. Ltd., London.)

saturator are not absorbed therein, but escape with steam into the multitubular preheaters and condensers, and after heating the entering ammoniacal liquors in the tubes in the preheater as they pass on their way to the still, the cooled gases consists of H_2S, HCN, and CO_2, finally escape either (1) directly into a furnace, where they are burnt in order to destroy the poisonous HCN, the H_2S burning to SO_2; or (2) into an **absorption apparatus** using iron oxide purifiers (as in coal gas works for purifying coal gas (see "Industrial Chemistry," Vol. I.); or (3) where the H_2S is recovered as sulphur, and nitrogen recovered as Prussian blue, the gas is burnt in a limited supply of air according to the **Claus** process, whereby the H_2S burns, depositing S, which is thereby recovered.

However, in most works the gases are simply burnt, the products of combustion escaping, unutilised, up the chimney.

MANUFACTURE OF AMMONIUM SULPHATE FROM MOND GAS.[1]

Very similar is the production of ammonim sulphate from **Mond Gas**.

The coal is charged into a distributor and hopper from an elevator, and then falls into the "generators" where, under the influence of a mixture of air and steam, it is gasified in the manner described in the section on Gaseous Fuels, "Industrial Chemistry," Vol. II., the equations being :—

$$C + O_2 = CO_2. \qquad CO_2 + C = 2CO. \qquad H_2O + C = CO + H_2.$$

There is thus produced a gas consisting mainly of CO, admixed with some hydrogen. The N in the coal escapes in 70-80 per cent. as NH_3 with the issuing gas. The gases next stream through a series of "coolers," in which they exchange their heat with the air which is streaming into the generators, the air being thus preheated, and the Mond gas cooled. Next, the gas passes through a washer, where it is largely freed from tar and dust. Finally, the gas passes up a tower, where it meets a stream of descending sulphuric acid, which combines with all the ammonia in the gas, forming ammonium sulphate. This acid solution is pumped by an acid pump up a tower, a continual circulation of the acid in the tower being kept up until the acid is practically saturated with ammonium sulphate, when it is run off and evaporated in a special apparatus, and the ammonium sulphate recovered. From the tower the gases pass through another tower, where they are treated with a stream of cold water, whereby the gas is cooled (and the water heated), and passes away directly to the gas engines or furnaces for use. The hot water obtained from the second tower is now pumped up the third tower, where it is allowed to flow down against an incoming current of cold air, which it thus saturates with water vapour, the air current then going on to the furnaces for the production of the CO by partial combustion of the coal.

While in coal gas works and coke ovens scarcely ever more than 20 per cent. of the N in the coal is obtained in the form of ammonia, in the Mond-gas process no less than 70-80 per cent. of the nitrogen present in the coal is ultimately converted into ammonia, and recovered as ammonium sulphate, *e.g.*, each ton of coal yields over 40 kg. of ammonium sulphate, against 10 kg. obtained in coke ovens. Over 4s. per ton profit can be made out of the ammonium sulphate thus recovered, which leads to a further reduction in the price of the gas for power.

According to Caro, even the waste obtained by washing certain coals—containing only 30-40 per cent. C. and 60-70 per cent. ash, and so useless for burning in the ordinary way—can be gasified by the Mond process, and a very considerable percentage of the nitrogen recovered as ammonium sulphate, 1 ton of this waste material yielding 25-30 kg. of ammonium sulphate and 50-100 H.P.-hours in the form of electrical energy.

Also **moist peat** can be gasified in the generators, and the contained nitrogen recovered as ammonium sulphate (Woltereck).

For further details the reader should see "Industrial Chemistry," Vol. II., "Gaseous Fuels"; also Caro, "Die Stickstoffrage in Deutschland" (1908); *Chemiker Zeitung.*, 1911, 505; Frank, Caro, and Mond, *Zeit. angew. Chem.*, 1906, 1569; Norton, *Consular Report*, pp. 40, 170; Lange, "Coal Tar and Ammonia," 1913, 861; Woltereck, English Patents, 16,504/04; 28,963/06; 28,964/06.

---

[1] See Martin's "Industrial Chemistry," Vol II., under Gaseous Fuels, where the composition of the gas and the nature of the furnaces, fuels, etc., are discussed.

MANUFACTURE OF AMMONIUM SULPHATE BY THE DIRECT PROCESS FROM COKE-OVEN GAS, BLAST FURNACE GAS, PRODUCER GAS, AND SIMILAR GASES RICH IN AMMONIA

This is a problem on which a great deal of elaborate work has been expended during the last fifteen to twenty years, and even at the present time it does not seem to have been completely solved.

Ammonium carbonate is very volatile, and consequently passing the gases directly through water only causes the formation of a dilute ammoniacal " gas liquor " (as in gasworks), the direct distillation of which would be expensive on account of fuel consumed.

It is, therefore, much more economical to pass the gases directly into fairly concentrated sulphuric acid, whereby matters must be so arranged that solid ammonium sulphate separates directly, that the concentrated sulphuric acid is not greatly diluted by the steam, etc., in the gases driven through it, and finally, that the resulting tar is not spoiled by the treatment.

One of the most successful systems is that embodied by the **Kopper Ammonia Recovery Plant**. The gases coming from the coke ovens, etc., are first cooled

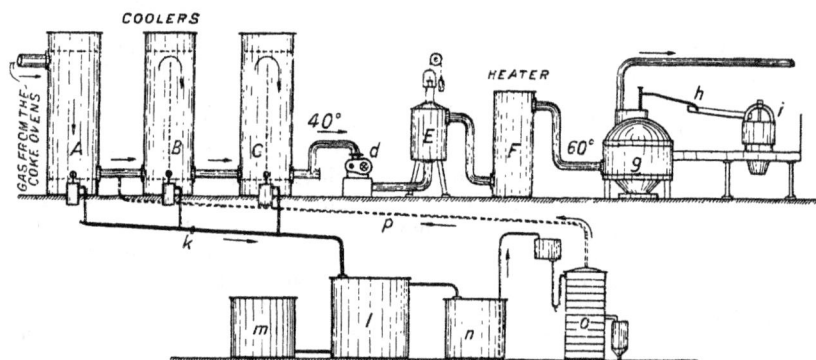

Fig. 14.—The Kopper Ammonia Recovery Plant.

to 40° C., whereby the heavy tar oils and the bulk of the steam (with 20-25 per cent. of the total ammonia) is deposited in the liquid form.

The gases are next passed through tubes whereby they are reheated to 60°-80° C.— the hot furnace gases in counter-current being used for this purpose—and the hot gases are then directly passed into 60° Bé_4, 141° Tw., H_2SO_4; simultaneously the NH_3 which has been expelled from the condensed liquors by heating them with lime is also passed into the H_2SO_4. The NH_3 directly combines with the H_2SO_4 and ammonium sulphate separates in a solid form in the saturator and is withdrawn from time to time, centrifuged, and dried as previously described.

Fig. 14 shows **Kopper's Plant**.

The hot furnace or coke-oven gases are passed through coolers *a*, *b*, *c*, until they are cooled to about 30°-40° C., and then are led through a Pelouze tar separator (Martin's "Industrial Chemistry," Vol. I.) *e*, so that almost all the tar (but not the light oils) are condensed, together with most of the water, which contains 25-75 per cent. of the total ammonia (according to the temperature), all the fixed ammonium salts (NH_4Cl, etc.), etc., etc. The tar and the ammoniacal water flows into a holder *l*, the tar being run off at the bottom into another holder *m*, while the ammoniacal water runs into a second holder *n*.

The tar-free gases are again reheated in *f* to 60°-80° C., and then are passed directly into the holder *g*, containing concentrated 60° Bé. (141° Tw.) sulphuric acid. The solid ammonium sulphate immediately separates, and is forced out from time to time by means of compressed air into the collecting tray *h*, thence into the centrifugal machine *i*, where excess of acid is drained off.

The reheating of the gas, before passing into the sulphuric acid, aids the evaporation of the water from the sulphuric acid. The heat of the reaction going on in g is alone often sufficient to achieve this, especially if all the fixed ammonium salts have been previously removed. The gases escaping from g contain benzene, and are passed on through heavy oils to extract this substance, as described under **Coke Ovens** in Martin's "Industrial Chemistry," Vol. II.

The ammoniacal water in n is distilled in a column with lime in the usual manner for ammonia (see p. 39), and the gaseous ammonia evolved is often directly sent back into the gas stream (between the coolers a and b) to be fixed by the sulphuric acid in g. Since the amount of deposited gas water here only amounts to about 20 per cent. of the washing water formerly needed to extract the ammonia in scrubbers, it is stated that the cost of the distillation of the ammonia and the quantity of the troublesome waste water is much less than by the ordinary process of the ammonia extraction.

In the **Otto-Hilgenstock Ammonia Recovery Process** (Fig. 15) the old condensing plant is entirely dispensed with, the tar being removed from the entering gases by a tar spray at A at a temperature above the dew-point of the liquors.

After depositing the tar in B the gases pass directly through an exhauster E into the saturator F, where the whole of the ammonia is caught by the sulphuric acid. The gases coming from the saturator are hot, and contain all their moisture in the form of steam; the gases are, therefore, passed forward to the oven flues, and the troublesome and offensive waste liquors are thus got rid of. C is the tar-spray pump, D the tar-spray feed pipe, G is the acid-spray catch box, H the mother liquor return pipe, J the tar store, K the tar-spray overflow pipe, L the condensing tank, M the pump delivering tar to railway trucks, N the pump delivering condensers to the saturator F.

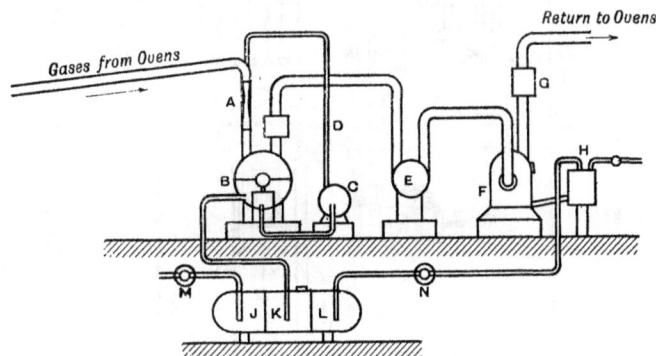

FIG. 15.—The Otto-Hilgenstock Ammonia Recovery Plant.

This process, by abolishing condensing plant, liquor tanks, ammonia stills, lime mixers, pumps, etc., effects a great saving, since less floor space is required, and nearly the whole of the steam required to distil the ammoniacal liquor made by the condensing process is abolished. Also no ammonia is lost, as often arises in a distilling plant. The ammonium sulphate produced contains 25-25.5 per cent. N and contains less than 0.1 per cent. of tar.

The **Coppée Company's Process of Semi-Direct Sulphate Recovery** is illustrated in Figs. 16 and 17.

MANUFACTURE OF CAUSTIC AMMONIA (LIQUOR AMMONIA)

A **crude** aqueous solution of ammonia, containing ammonium sulphide and sometimes carbonate, is manufactured under the name "**Concentrated Ammonia Water**" by distilling ordinary gas-water without addition of lime in column apparatus.

Concentrated Ammonia Water is used as a convenient source of ammonia either for ammonia-soda factories, or for factories which, possessing no ammonia plant of their own, require ammonia gas in a concentrated form, *e.g.*, for manufacturing salts such as NH_4Cl, NH_4NO_3, etc., or liquid NH_3. It contains much less CO_2 and H_2S than the ordinary diluted ammoniacal liquors of gas-works, because these gases, in general, escape during the distillation before the NH_3 gas. This liquor is much cheaper than the pure aqueous solution, and the presence of some CO_2 and H_2S does not affect the manufacture of many chemicals.

DIAGRAM OF BYE PRODUCT PLANT WITH SEMI DIRECT SULPHATE RECOVERY

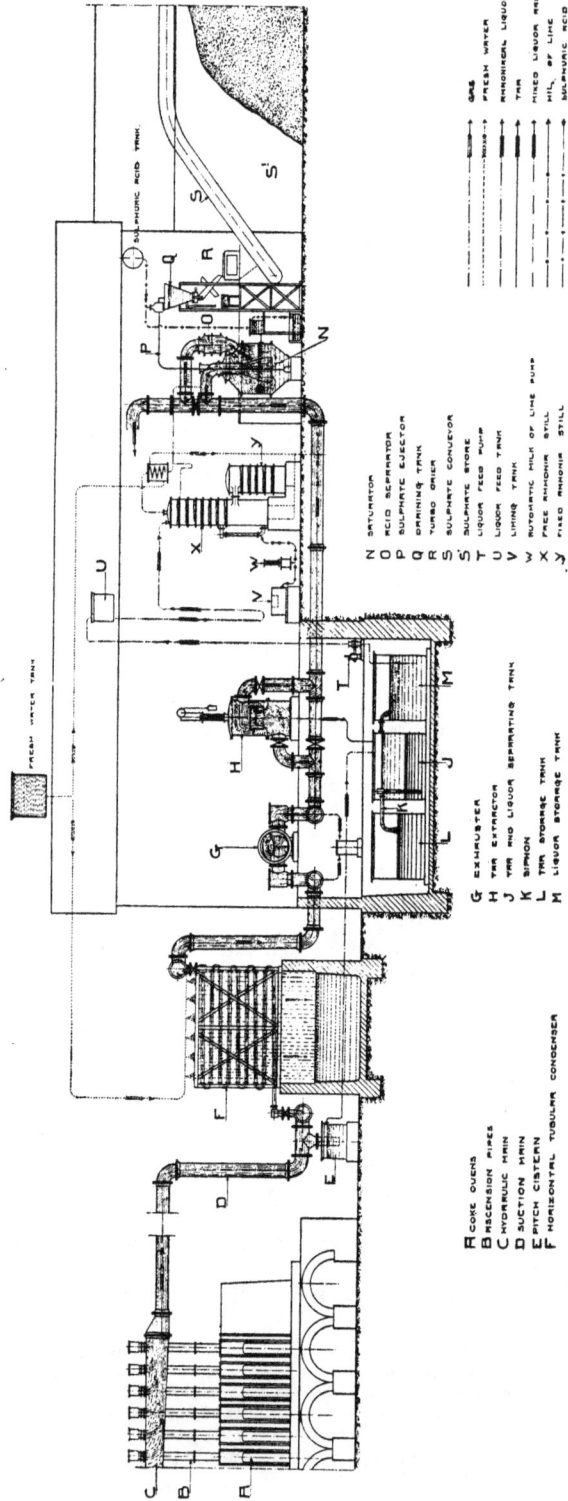

A COKE OVENS
B ASCENSION PIPES
C HYDRAULIC MAIN
D SUCTION MAIN
E PITCH CISTERN
F HORIZONTAL TUBULAR CONDENSER

G EXHAUSTER
H TAR EXTRACTOR
J TAR AND LIQUOR SEPARATING TANK
K SIPHON
L TAR STORAGE TANK
M LIQUOR STORAGE TANK

N SATURATOR
O ACID SEPARATOR
P SULPHATE EJECTOR
Q DRAINING TANK
R TURBO DRIER
S SULPHATE CONVEYOR
S' SULPHATE STORE
T LIQUOR FEED PUMP
U LIQUOR FEED TANK
V LIMING TANK
W AUTOMATIC MILK OF LIME PUMP
X FREE AMMONIA STILL
Y FIXED AMMONIA STILL

GAS
FRESH WATER
AMMONIACAL LIQUOR
TAR
MIXED LIQUOR AND TAR
MILK OF LIME
SULPHURIC ACID

THE COPPÉE COMPANY (GREAT BRITAIN) LTD., KING'S HOUSE, KINGSWAY, W.C.

FIG. 16.—Bye-Product Plant with Semi-direct Sulphate Recovery.

46

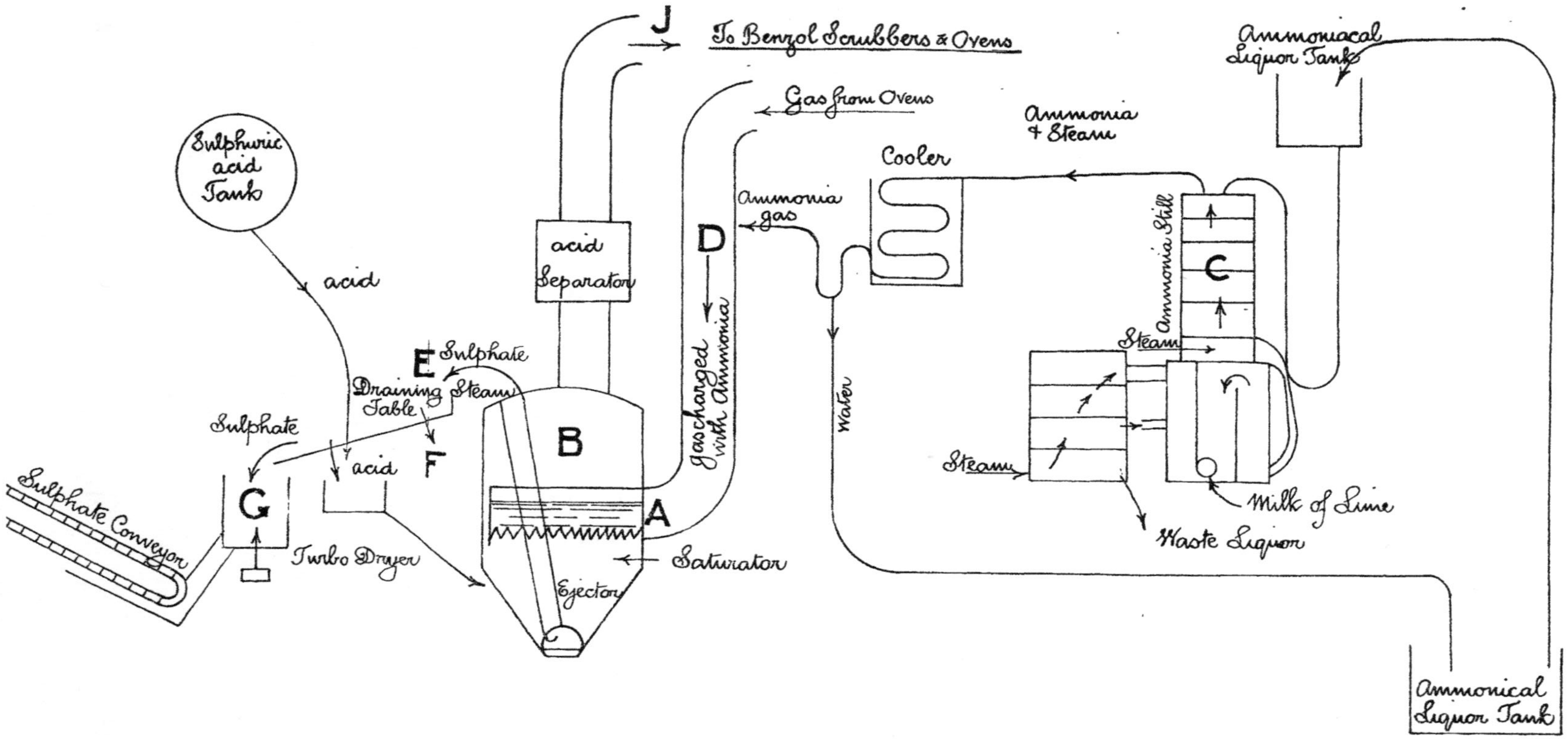

Sulphate Plant for Direct Recovery Process

Ammoniacal Liquor Tank

Ammoniacal Liquor Tank

Milk of Lime

Waste Liquor

Ammonia Still

C

Steam

Steam

Steam

Ammonia + Steam

Cooler

Gas from Oven

To Benzol Scrubbers & Ovens

Ammonia gas

Water

Discharged with Ammonia

D

Saturator

A

acid Separator

B

Ejector

J

Sulphate

E

Draining Steam Table

F

acid

Sulphuric acid Tank

acid

Sulphate

Turbo Dryer

G

Sulphate Conveyor

FIG. 17.

(By Permission of the Coppée Co. Ltd., London.)

To face page 47.

Two varieties of "concentrated ammonia water" are manufactured :—(1) One containing 15-18 per cent. of NH_3 with both sulphide and carbonate present; (2) one containing 18-26 per cent. NH_3 containing some sulphide but practically no carbonate.

(1) The first liquid is made by passing the ammoniacal gases from the still (as described above) through a reflux condenser, whereby some moisture is removed, and then into a direct condenser, the gases from which are washed through water. The formation of ammonia carbonate and resulting blockage of the pipes prevents a higher concentration of NH_3 being obtained in this manner.

(2) To manufacture the second liquid we proceed as before, but the vapours from the still are passed first through the reflux condenser and then through vessels containing milk of lime which removes the CO_2 and some H_2S. The vapours are then condensed and contain 22-26 per cent. NH_3, some H_2S, but little or no CO_2.

Manufacture of Pure Aqueous Solutions of NH_3.—The ammoniacal liquor is first heated to 70°-80° C. in order to expel most of the CO_2 and H_2S. The liquid is next distilled in a columnar apparatus, as previously described, and the vapours are passed, first through a reflux condenser (to remove some water), then through milk of lime washers to remove CO_2 and H_2S (the partly used lime being run back into the still to decompose the fixed ammonium salts), the last traces of H_2S being removed by either passing the gases through $FeSO_4$ solution, or through a little NaOH solution. J. Louis Foucar recommends ammonium persulphate or sodium permanganate.

Lastly, the vapours pass through wood-charcoal, which removes tarry matters, and then (sometimes) through a non-volatile fatty or mineral oil. The fairly pure NH_3 gas is then led into distilled water until the required concentration (up to about 36 per cent. NH_3 is attainable) is obtained.

The charcoal filters are revivified from time to time by ignition in closed retorts.

The following table gives the specific gravity of aqueous ammoniacal solutions of various strengths (after Lunge and Wiernek) :—

SPECIFIC GRAVITY OF AQUEOUS SOLUTIONS OF AMMONIA AT 15° C.

Specific Gravity.	NH_3 Per Cent.	1 Litre Contains g. HN_3.	Specific Gravity.	NH_3 Per Cent.	1 Litre Contains g. NH_3.
1.000	0.00	0.0	0.936	16.82	157.4
0.998	0.45	4.5	0.934	17.42	162.7
0.996	0.91	9.1	0.932	18.03	168.1
0.994	1.37	13.6	0.930	18.64	173.4
0.992	1.84	18.2	0.928	19.25	178.6
0.990	2.31	22.9	0.9.6	19.87	184.2
0.988	2.80	27.7	0.924	20.49	189.3
0.986	3.30	32.5	0.922	21.12	194.7
0.984	3.80	37.4	0.920	21.75	200.1
0.982	4.30	42.2	0.918	22.39	205.6
0.978	5.30	51.8	0.916	23.03	210.9
0.974	6.30	61.4	0.914	23.68	216.3
0.972	6.80	66.1	0.912	24.33	221.9
0.968	7.82	75.7	0.910	24.99	227.4
0.966	8.33	80.5	0.908	25.65	232.9
0.964	8.84	85.2	0.906	26.31	238.3
0.962	9.35	89.9	0.904	26.98	243.9
0.960	9.91	95.1	0.902	27.65	249.4
0.958	10.47	100.3	0.900	28.33	255.0
0.956	11.03	105.4	0.898	29.01	260.5
0.954	11.60	110.7	0.896	29.69	266.0
0.952	12.17	115.9	0.894	30.37	271.5
0.950	12.74	121.0	0.892	31.05	277.0
0.948	13.31	126.2	0.890	31.75	282.6
0.946	13.88	131.3	0.888	32.50	288.6
0.944	14.46	136.5	0.886	33.25	294.6
0.942	15.04	141.7	0.884	34.10	301 4
0.940	15.63	146.9	0.882	34.95	308.3
0.938	16.22	152.1	...	...	...

TECHNICAL AMMONIUM SALTS

Ammonium Sulphate $(NH_4)_2SO_4$.—Manufacture and properties are described on p. 38 *et seq.* See also under **Manures**, "Industrial Chemistry," Vol. II.

Heat of formation is given by :—

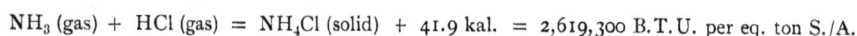

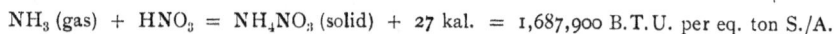

$$NH_3 \text{ (gas)} + \tfrac{1}{2}H_2SO_4 = \tfrac{1}{2}(NH_4)_2SO_4 \text{ (solid)} + 25.5 \text{ Cal.} = 1,594,100 \text{ B.T.U. per eq. ton S./A.}$$

Ammonium Chloride, (Sal Ammoniac), NH_4Cl.—Prepared by passing vapours of ammonium into HCl, or *vice versa*. Sal ammoniac is the sublimed chloride. It is somewhat expensive to sublime owing to the difficulty of obtaining suitable vessels into which to sublime the chloride. Cheap earthenware is often used, which, however, can only be used once, as it has to be broken to remove the sublimate of sal ammoniac.

The substance is used in galvanising, in soldering, in galvanic cells, in the manufacture of colours, in calico-printing, in pharmacy.

Heat of formation is given by :—

$$NH_3 \text{ (gas)} + HCl \text{ (gas)} = NH_4Cl \text{ (solid)} + 41.9 \text{ kal.} = 2,619,300 \text{ B.T.U. per eq. ton S./A.}$$

Ammonium Carbonate.—Commercial ammonium carbonate is usually a mixture of ammonium bicarbonate NH_4HCO_3, and ammonium carbonate $NH_4O.CO.NH_2$.

The ammonia content of the mixture varies between 25-58 per cent., the usual percentage of ammonia being 31 per cent.

It is most easily made by direct combination of ammonia, carbon dioxide, and water vapour, the substance being condensed on water-cooled surfaces of aluminium The substance is volatile, and should at once be packed into air-tight vessels in order to avoid loss.

It is used in the manufacture of baking powders, dyeing, in extracting colours from lichens, in caramel making, smelling salts, etc. Also as a general detergent, and for removing grease from fabrics.

Ammonium Nitrate, NH_4NO_3.—See p. 13 for manufacture and properties. Its main use is for safety explosives, sporting powders, fireworks, etc. Much is used for the preparation of nitrous oxide, "laughing gas."

Heat of formation is given by :—

$$NH_3 \text{ (gas)} + HNO_3 = NH_4NO_3 \text{ (solid)} + 27 \text{ kal.} = 1,687,900 \text{ B.T.U. per eq. ton S./A.}$$

Ammonium Perchlorate, NH_4ClO_4 is prepared by the double decomposition of $NaClO_4$ and NH_4Cl.

For Patents bearing on its manufacture see **Alvisi**, D.R.P., 103,993; **Miolati**, D.R.P., 112,682; see also **Witt**, *Chem. Ztg.*, 1910, p. 634.

The salt is obtaining extended use as an explosive and oxidising agent, soluble in 5 parts cold water, insoluble in alcohol, sp. gr. 1.89.

Ammonium Phosphate, $(NH_4)_2HNO_4$, is made by neutralising phosphoric acid with ammonia, evaporating, and crystallising. Used in the manufacture of sugar (Lagrange process), and in the impregnation of matches.

Ammonium Persulphate, $(NH_4)_2S_2O_8$, prepared by electrolysing an acid saturated solution of ammonium sulphate at 7° C., using 5 volts in a specially designed apparatus. Much used as an oxidising agent in dyeing and photography ; see the D.R.P. 155,805, 170,311, and 173,977 ; also Marshall, *Trans. Chem. Soc.,* 1891, p. 777.

Ammonium Thiosulphate $(NH_4)_2S_2O_3$, prepared by double decomposition, thus—$2NH_4Cl + Na_2S_2O_3 = (NH_4)_2S_2O_3 + 2NaCl$. Used in photography.

Ammonium Acetate, $NH_4OOC.CH_3$, prepared by neutralising acetic acid with ammonia, and used for making acetamide.

Ammonium Fluoride, NH_4F, made by neutralising HF with NH_3, is used for etching glass, decomposing minerals containing rare earths, for the manufacture of incandescent mantles, for preparing antimony fluoride and other technically important metallic fluorides, and also, to some extent, in dyeing.

Ammonium Sulphocyanide, NH_4CNS, occurs in gas liquor, and is often prepared by adding flowers of sulphur to an ammoniacal coal-gas washer, ammonium polysulphide being formed and cyanogen absorbed. It can also be made from CS_2 and NH_3 :— $CS_2 + 2NH_3 = NH_4CNS + H_2S$. Used in photography, calico-printing, and dyeing.

Ammonium Chlorate, NH_4ClO_3, used in fireworks and explosives.

Ammonium Bromide, NH_4Br, used in pharmacy and photography.

Ammonium Oleate is used in ammonia soaps.

Dry Ammonia, $CaCl_28NH_3$, containing 60 per cent. of NH_3, is made by direct combination, and has in Germany a market as a portable and compact form of ammonia.

"Solid Ammonia" is manufactured by the Chemische Fabrik Betterhausen, Marquart & Schultz, by adding to a mixture of 3-5 parts of sodium stearate (dissolved in 10 parts of aqueous ammonia or 80 per cent. spirits of wine) about 85 or 90 parts of ammonia solutions containing 25-33 per cent. of ammonia. The mixture has a consistency nearly equal to that of solid paraffin. When heated or left exposed to air it gives up the whole of its ammonia, leaving behind the solid sodium stearate.

Large quantities of ammoniated superphosphates are made in the United States of America, containing up to 6 per cent. NH_3, and made by treating the superphosphate with ammonia or merely by mixing in ammonium sulphate into the superphosphate. Products are used for manurial purposes.

Anhydrous Ammonia (Liquid Ammonia) is simply the purified NH_3 gas liquefied under pressure and filled into steel cylinders.

It is the most suitable and efficient working substance for refrigerating machines, and is used in some wool-washing institutions for cleaning purposes, the substance being an excellent solvent.

The liquid boils at 34° C. At 15° C. the liquid has a vapour tension of 6 atmospheres. As a liquid it has a very large coefficient of expansion, and a specified weight at 15° C. of 0.614, at 60° C. of 0.540.

In order to manufacture the substance, excess of lime is added to crude gas water so as to fix all the CO_2 and H_2S, as well as to set free all the ammonia from the "fixed" ammonium salts. The liquid is then distilled in a special column apparatus, somewhat similar to that described on pp. 38 for ammonium sulphate, but somewhat more complicated, the large masses of mud-like lime requiring special vessels for mixing, depositing the precipitated mud, and for boiling out the ammonia. The NH_3 gas emerging from the columns is cooled and then made to traverse a number of vessels containing milk of lime, whereby the last traces of CO_2, H_2S, etc., are removed. The vapour then passes through a layer of paraffin oil, which retains tar and pyridine, etc. Then the gas passes through charcoal filters to remove the last traces of tarry matters, etc. The dry and pure vapour next passes to the pumps, where it is liquefied under 8 atmospheres' pressure. The operation is usually carried out in two stages, the gas heated in the first compression pump being cooled by water cooling before being passed into the next pump where the final liquefaction takes place.

The liquid is stored in steel cylinders, usually made to hold either 20 kg. or 50 kg. of liquid ammonia. The cylinders should be tested every four years at 30 atmospheres' pressure. For each 1 kg. liquid ammonia there is allowed a volume of 1.86 l. The liquid ammonia, on evaporation, should not leave behind more than 0.1 per cent. residue, consisting of water, machine oil, pyridine, etc.

CHAPTER V

—

Synthetic Ammonia

CHAPTER V

SYNTHETIC AMMONIA

LITERATURE

F. Haber and Le Rossignol.—"Technical Preparation of Ammonia from its Elements," *Zeit. für Elektrochemie*, 1913, **19**, 53, 72. (There is a good abstract of this paper in the *Journ. Soc. Chem. Ind.*, 1913.) *Ber.*, 1907, **40**, 2144; *Zeit. Elek.*, 1908, **14**, 181, 513.

Haber and van Oordt.—*Zeit. Anorg. Chem.*, 1905, **43**, 111; 1905, **44**, 341.

Bernthsen.—"Eighth International Congress of Applied Chemistry." New York, 1912. Abstract, *Journ. Soc. Chem. Ind.*, 1912, **31**, 982.

Donath and Indra.—"Die Oxydation des Ammoniaks zu Saltpetersäure und Saltpetriger Säure," pp. 54-67. Stuttgart, 1913.

Lunge.—"Coal Tar and Ammonia." Fourth Edition, pt. 2, p. 815.

Knox.—"The Fixation of Atmospheric Nitrogen." 1914.

Norton.—"Utilisation of Atmospheric Nitrogen." 1912.

Caro.—"Die Stickstoffrage in Deutschland." 1908. *Zeitsch. angew. Chem.*, 1910, **23**, 2412.

Serpek.—English Patent, 13,086, 1910.

J. W. Richards.—*Trans. Amer. Electrochem. Soc.*, 1913, **23**, 351.

Fraenkel.—*Zeitsch. Elektrochem.*, 1913, **19**, 362.

S. A. Tucker.—*Journ. Soc. Chem. Ind.*, 1913, **32**, 1143; *Journ. Ind. and Eng. Chem.*, 1913, **5**, 191.

Also references and patents given in the text.

Several methods of making atmospheric nitrogen unite to form ammonia have been proposed, and have been commercially successful. In the course of a few years large amounts of synthetic ammonia will be on the market produced by one or other of these methods.

(1) AMMONIA BY DIRECT UNION OF NITROGEN AND HYDROGEN BY MEANS OF A CATALYST

LITERATURE

F. Haber and R. Le Rossignol.—"Technical Preparation of Ammonia from its Elements," *Zeitschrift für Elektrochemie*, 1913, **19**, 53-72.

Under certain conditions nitrogen directly unites with hydrogen to form ammonia according to the equation:—

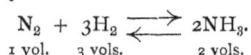

$$N_2 + 3H_2 \rightleftharpoons 2NH_3.$$

1 vol.   3 vols.   2 vols.

Heat evolved according to the equation:—

$$N + 3H = NH_3 \text{ (gas)} + 11,900 \text{ calories} = 727,000 \text{ B.T.U. per eq. ton S./A.}$$

This equation is reversible, depending upon the equilibrium represented by the expression:—

$$K = \frac{P_{NH_3}}{P_{N_2}^{\frac{1}{2}} P_{H_2}^{\frac{3}{2}}},$$

where P_{NH_3}, P_{N_2}, P_{H_2} respresent the partial pressures of the respective gases, NH_3, N_2, H_2, and K is the reaction constant.

It will be noticed that 4 vols. of the mixture of nitrogen and hydrogen produce 2 vols. of ammonia, and consequently, as in the case of all gaseous reactions where the products of interaction occupy a smaller volume than the original components, and increase of pressure favours the formation of the products possessing the least volume. Haber and his co-workers found that by employing a very high pressure, about 200 atmospheres, and a temperature of between 500°-700° C., and by passing the mixture of gases over a catalyst, such as osmium or uranium, the combination of nitrogen with hydrogen proceeded so favourably that from 3-12 per cent. of ammonia was formed in the reacting gases.

It will be seen from the above equation that large amounts of energy are not required for the production of ammonia from its elements, and that, therefore, the manufacture of synthetic ammonia need not be confined to districts where large amounts of cheap water-power are available, as is the case with the electrical production of nitric acid, cyanamide, etc.

The Badische Anilin und Soda Fabrik have now erected works at Oppan, near Ludwigshafen, for the large scale preparation of synthetic ammonia by this process.

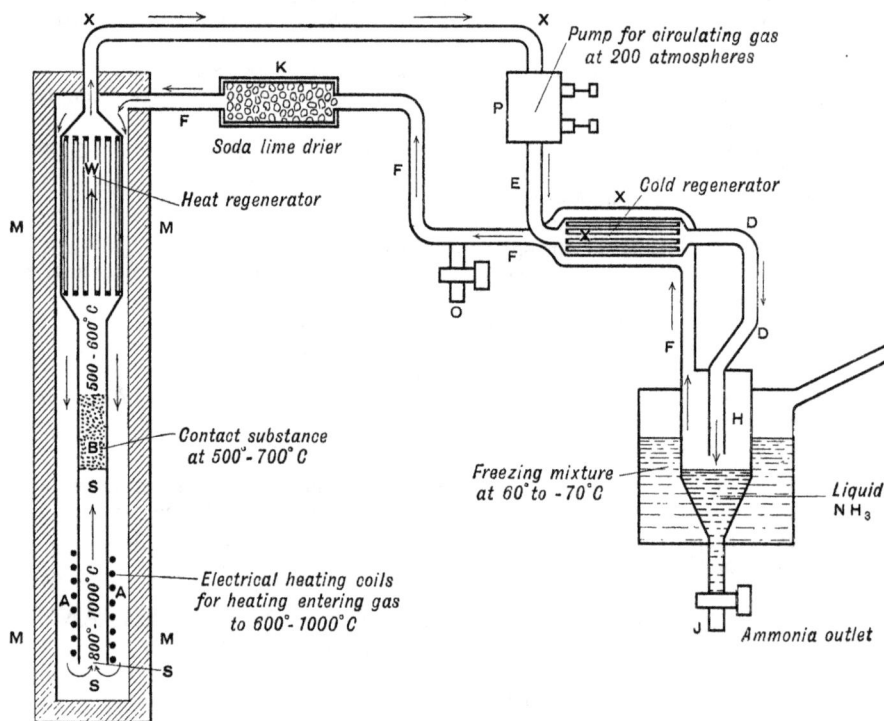

FIG. 18.—The Haber and Le Rossignol Process for Synthetic Ammonia.

Although the details of the plant employed on the large scale have not been published, the details of the experimental plant employed by Haber and Le Rossignol have been described in the above publication.

A diagrammatic sketch of Haber and Le Rossignol's experimental apparatus is given in Fig. 18.
Through the tube F a mixture of 1 vol. nitrogen and 3 vols. of hydrogen under a pressure of 200 atmospheres enters the strong steel vessel MM. After passing over the outer surface of a number of capillary metallic tubes W—which serve as a heat interchanger and regenerator as we will presently explain—the gas passes down the tube as shown, over the surface of an electrical heating coil AA, where the temperature of the gas is increased to 800°-1,000° C., then back up an interior iron tube SS, over the layer of catalytic substance B, thence through a number of capillary tubes W, out through the tube XX, thence through the compressing pump P, working at 200 atmospheres' pressure, thence out through the tube E, through the set of capillary tubes X, and so into the vessel H, which is surrounded by a freezing mixture of solid CO_2 and ether at a temperature of −60° or −70°, which causes the ammonia in the gas to separate in the liquid state, whence it can be drawn off by the cock J. From H the gases pass away by the tube F, over the exterior surface of the system

of capillary tubes x, thence after passing over a soda-lime drier κ, the gas enters M as previously described.

The mode of action of the apparatus is as follows :—The cold gas entering MM by the pipe F is heated by passing over the bundle of capillary tubes W conveying the hot gas away from the contact substance B.   Thus the entering gas is, by the time it has left W, preheated to a temperature of 400°-500° C., and in so doing has abstracted practically all the excess heat from the hot gas passing away from B, so that this latter, by the time it reaches the pipe XX, is practically at atmospheric temperature, while at the same time the entering gas, by the time it reaches the heating coil A, is already at a high temperature, so that practically no loss of heat occurs.   For this reason W is called the "heat regenerator."   The hot gas thus entering the tube SS is further heated in its passage by the electrical heating coils surrounding the end of the tube SS to a temperature of 800°-1000° C.   The hot gas then passes into the contact substance B, which is maintained by the hot gas at a temperature of 500°-700° C.   Here the formation of ammonia takes place, 3-7 per cent. of the entering nitrogen and hydrogen escaping as NH_3, along with excess of uncombined nitrogen and hydrogen.   The hot mixture of gases from B then streams through the series of fine capillary tubes W, and in so doing gives up practically all its heat to the cold entering stream of gas coming into the apparatus at F.   The gaseous mixture, now cooled to ordinary temperatures, passes away through the pipe XX into the pump P, working at 200 atmospheres, and then passes through the series of metallic capillary tubes X.   While passing through these it meets with a cold stream of gas at −60° C. coming from the vessel H.   Consequently the gas in the capillary tubes XX parts with its heat to the cold gas coming from H, being itself chilled in so doing, and passes out of X through the pipe DD into H at a temperature only slightly above that of the cold gas escaping from H.   The cold gas passing from H up F is heated almost up to atmospheric temperature by the capillary tubes X, and thus passes away through the drier κ and enters MM at F at a temperature only very slightly below atmospheric.   For this reason the tubes X are called the "cold regenerator."

The gas entering H contains 3-7 per cent. of NH_3, and this condenses in a liquid form at the low temperature (−60° C. to −70° C.) prevailing therein, owing to the surrounding freezing mixture of solid CO_2 and ether.   This ammonia can be drawn off in a liquid form by the tap J, or, if required in a gaseous form, can be so obtained by opening the outlet valve to a suitable extent.

As the ammonia is withdrawn a fresh supply of nitrogen and hydrogen is added through the valve O, so that the operation is a continuous one.

In practice, the very serious engineering operations of working continuously a plant with gas at 200 atmospheres, without leakage, has been got over by carefully turned screw joints, one part of an angle of 16° C. screwing into another part of an angle of 20°, so that perfectly gas-tight connections were made in this way.   For special details of construction the original memoirs should be consulted.

Haber and his co-workers have made numerous experiments on the most suitable catalysts to use as contact substances, describing the results of experiments with cerium and allied metals, manganese, tungsten, uranium, ruthenium, and osmium.   The best catalyst proved to be finely divided osmium, but as this substance is limited in quantities and very expensive, it was found that uranium (pieces the size of a pin's head) also acted efficiently.

Thus, in one series of experiments, commercial uranium, broken up with a hammer, was used in a column 4·5 mm. diameter and 3-3.5 cm. long.   At 600° C. a vigorous formation of ammonia took place.   At 190 atmospheres and with the gas mixture passing through the apparatus at 20 l. per second (measured at atmospheric pressure and temperature) it issued with an ammonia contents of 5.8 per cent. by volume.

The nitrogen can be obtained from the atmosphere, either by liquefying it and fractionally distilling it with a Linde or Claude machine, as described in Martin's "Industrial Chemistry," Vol. II., or simply by passing air over heated copper.   The hydrogen can be obtained industrially by any of the methods discussed in Martin's "Industrial Chemistry," Vol. II.

In 1913 ammonia in the form of commercial 25 per cent. (NH_3) ammonium sulphate possesses a value of 4.75d. per lb. (89 Pf. per kilo), while the nitrogen and hydrogen contained therein may be valued at 1.07d. per lb. (2½ Pf. and 17½ Pf. for the H and N respectively, in 1 kilo of ammonia).

It has been stated that the total costs of manufacturing ammonium sulphate from this synthetic ammonia only amounts to £2. 6s. 6d. per ton, and so the process appears to be the one which has a greater chance of success as regards low cost than any other method yet proposed, especially as the plant can be erected anywhere (see p. 28).

Other Processes for Making Synthetic Ammonia from Atmospheric Nitrogen and Hydrogen.—De Lambilly (German Patents, 74,274 and 78,573) passes a mixture of N_2, steam, H_2 and CO over Pt (or other catalyst) at 80°-130° C., when ammonium formate is formed ($N_2 + 3H_2 + 2CO + 2H_2O = H.COONH_4$).   With CO_2 at 40°-60° C. ammonium bicarbonate is formed ($N_2 + 3H_2 + 2CO_2 + 2H_2O = 2NH_4HCO_3$).

Schlutius (English Patent, 2,200/02) passes Dowson gas (39 per cent. CO, 4 per cent. CO_2, 43 per cent. N_2, 14 per cent. H_2) and steam over Pt in the presence of a silent electrical discharge.   Below 80° NH_3 is produced, above 80° C. ammonium formate.

The reader may also see the patents :—Young, English Patent, 1,700/80 ; Hooper, U.S. Patent, 791,194 ; Cassel, German Patent, 175,480 ; Gorianoff, French Patent, 368,585 ; see also Davies, *Zeit. physical Chem.*, **64**, 657 ; Briner and Mettler, *C. R.*, 144, 694 ; Donath and Indra (*loc. cit.*, pp. 54-67) and Lunge, "Coal Tar and Ammonia," Fourth Edition, pt. 2, p. 815 give details of other processes.

(2) Ammonia from Cyanamide.—Cyanamide is made by causing atmospheric nitrogen to directly unite with calcium carbide, as described in Martin's "Industrial Chemistry," Vol. II. Ammonia is next made from the cyanamide by passing superheated steam over it, when the following change takes place :—

$$CaNCN + H_2O = CaCO_3 + 2NH_3.$$

The operation may be carried out as indicated in Fig. 19.

Superheated steam is led in through the pipe A into the chamber B filled with the cyanamide on trays as indicated.

Ammonia is here generated, which may be directly drawn off or else led into acid and fixed.

The crude calcium cyanamide contains, it will be remembered, much carbon in the form of **graphite**, the substance as put on the market having the approximate composition CaNCN + C.

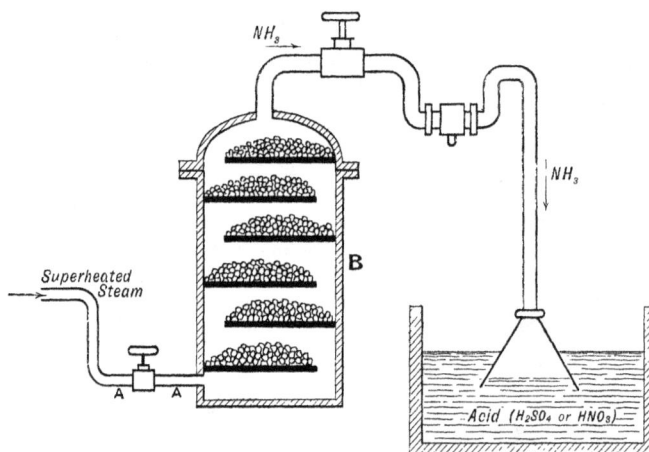

FIG. 19.—Ammonia from Cyanamide.

After treating with steam, as above described, we have the residue of $CaCO_3 + C$ left. This may be either worked for graphite (by dissolving the $CaCO_3$ in acid which leaves the C as graphite unaffected), or the residue may be returned to the lime furnaces, and be there calcined for the production of CaO, the lime thus produced containing one-third the proper amount of free carbon necessary for the production of calcium carbide when returned to the calcium carbide factory ($CaCO_3 + C = CaO + C + CO_2$ and $CaO + 3C = CaC_2 + CO$).

According to J. Louis Foucar, since cyanamide has to be made from calcium carbide and nitrogen, and the calcium carbide in its turn from lime and a plentiful supply of electrical energy, the ammonia made by this process could not, theoretically, be produced as cheaply as the direct synthetic ammonia prepared by direct union of nitrogen and hydrogen. Foucar (private communication) worked out the costs of manufacture of ammonium sulphate from cyanamide as follows :—

	£	s	d	
Cost of carbide - - - - -	3	5	0	per ton.
Cost of nitrogen - - - - -	0	2	0	
Cost of steam - - - - -	0	0	6	
Cost of sulphuric acid - - - -	0	16	0	
Less value of graphite - - - -	0	5	6	
Total net cost for materials - -	3	18	0	per ton
Cost for powder (cyanamide only) - - -	0	7	6	
Labour (cyanamide, steaming, etc.), power, repairs, depreciation, rent, taxes, amortisation, salaries of staff, and other charges - - -	0	7	6	
Total cost of manufacture - -	4	13	0	per ton.

The selling price of ammonium sulphate in England in 1912 was £14 per ton, and the total cost of manufacture from gas-liquor was £3 per ton.

(3) **Ammonia from Nitrides**.—A great many proposals have been made to use nitrides, either directly as fertilisers, or to produce ammonia therefrom by the action of superheated steam.

One process actually in use is the **Serpek Process**,[1] in which **aluminium nitride**, AlN, is used.

Bauxite (a naturally occurring impure hydrated alumina) is heated with coal in an atmosphere of nitrogen at a temperature of 1700°-1800° C. in a specially designed electric furnace, when N is bsorbed, thus :—

$$Al_2O_3 + 3C + N_2 = 2AlN + 3CO.$$

The absorption of N begins at 1100° C. ; at 1500° C. the absorption is rapid, the velocity increasing up to 1800°-1850° C., where almost violent absorption takes place, nearly chemically pure AlN being produced. Above 2000° the nitride decomposes.

The reaction is strongly endothermic, Fraenkel calculating the heat absorbed in the above reaction to be − 243,000 calories, and Richards, − 213,220 calories.

According to Richards (*loc. cit.*) two superimposed cylinders, A and B, Fig. 20 which rotate in opposite directions, are used. Powdered bauxite is sent into A at C

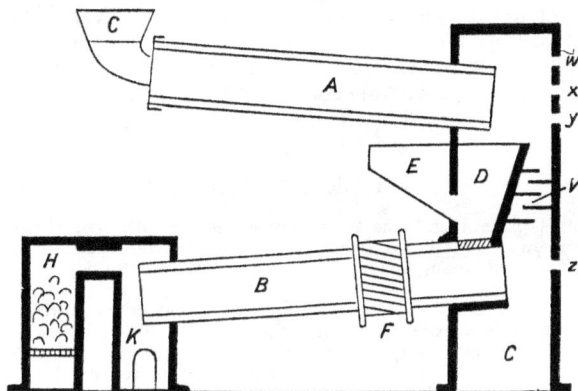

FIG. 20.—The Serpek Process.

and is calcined in its descent by the hot gases from the reaction going on in the lower cylinder B, and also by the combustion of the CO gas coming from the producer K and evolved in the reaction, this combustion being carried on in the side furnace c, air being admitted by the flues w, x, y, z, and the baffle plates v remove dust from the ascending gases. As the result of this the highly heated bauxite falls into a hopper D, and is there mixed with the requisite amount of carbon by means of the side hopper E. The bauxite and carbon fall into the lower rotating cylinder B (made of iron lined with compressed aluminium nitride), and thence passes into the electric resistance furnace F (made of a series of bars of compressed carbon and AlN, crossing the furnace diametrically or embedded longitudinally in the lining of the furnace), where the mixture is heated to 1800°-1900° C., and there meets the nitrogen from the producer-gas plant H (evolving 70 per cent. N_2 + 30 per cent. CO at 400° C.) placed at the lower end of the apparatus.

The nitrogen is absorbed, and the resulting aluminium nitride, in the form of a grey powder, passes on to an air-tight chamber K at the bottom of the apparatus.

[1] For a description of the process see Serpek's patent, English Patent, 13,086 of 1910. Also J. W. Richards, *Met. and Chem. Eng.*, 1913, **11**, 137 ; also *Trans. Amer. Electrochem. Soc.*, 1913, **23**, 35 ; S. A. Tucker, *Journ. Soc. Chem. Ind.*, 1913, **32**, 1143 ; *Journ. Ind. and Eng. Chem.*, 1913, **5**, 191 ; Fraenkel, *Zeitsch. Elektrochem.*, 1913, **19**, 362.

The silicious impurities in the charge are largely volatilised out ; consequently, using crude bauxite, the mass contains 26 per cent. N, but using pure alumina, Al_2O_3, 34 per cent. N can be obtained, corresponding to pure AlN.

The resulting aluminium nitride, AlN, is next decomposed by caustic soda to form ammonia and sodium aluminate :—

$$AlN + 3NaOH = NH_3 + Na_3AlO_3.$$

From the sodium aluminate thus produced, pure alumina can be obtained in the manner described in Martin's "Industrial Chemistry," Vol. II.

According to the Badische Anilin u. Soda Fabrik Patents (German Patents, 235,300, 235,765, 235,766, 235,868, 236,395) the elimination of ammonia by alkalies is facilitated by the addition of NaCl and other soluble salts. The decomposition of the AlN can be effected by acids, thus :—$2AlN + H_2SO_4 + 6H_2O = 2Al(OH)_3 + (NH_4)_2SO_4$. These patents work out methods for recovering the $Al(OH)_3$ in a pure state.

The same firm in the German Patent, 243,839, states that the formation of nitride from alumina and coal is greatly accelerated by the addition of 5-10 per cent. of the oxides of certain elements, which themselves yield stable nitrides, among which SiO_2 and the oxides of Ti, Zr, V, Be, Mo, Ur, Ce, Cr, also silicates, vandates, etc. Iron oxide does not influence the catalytic action of these oxides.

Aluminium nitride, as prepared by Serpek, is of value both for its combined nitrogen and also for the production of pure alumina for the manufacture of metallic aluminium. Without the sale afforded by this alumina it is doubtful whether the process would be economical, as the alumina would have to be used over and over again. As it stands, however, the alumina produced as a by-product finds a ready sale, and the ammonia simultaneously produced is also a valuable product.

The process is being worked by the Société Générale des Nitrates in France.

OTHER NITRIDES

Various suggestions have been made to use **boron nitride**, BN, but the high temperature required for its formation and the volatility of the resulting boric acid when the nitride is decomposed by superheated steam for ammonia ($2BN + 3H_2O = 2B(OH)_3 + 2NH_3$) have formed insuperable difficulties.

Calcium nitride, Ca_3N_2, containing 18.9 per cent. of N, and lithum nitride, Li_3N, containing 38.8 per cent. N, are easily prepared from the elements and easily decompose, giving ammonia. Magnesium nitride, Mg_3N_2, containing 27.7 per cent. N, is likewise easily prepared by heating magnesium in nitrogen gas. When heated in hydrogen it is converted into ammonia and hydride ($Mg_3N_2 + 6H_2 = 2NH_3 + 3MgH_2$).

However, none of the processes suggested to employ these reactions as technical sources of ammonia appear to be successful.

The attempts to utilise titanium nitride by heating it in a mixture of nitrogen and hydrogen have also been unsuccessful.

Silicon nitride, Si_3N_4, containing 42.5 per cent. N, has been suggested for direct use as a fertiliser, as giving results possibly superior to ammonia and cyanamide. The Badische Anilin und Soda Fabrik (English Patent, 16,368 of 1910) have patented processes for fixing atmospheric nitrogen by means of silicon nitride. They point out that in preparing silicon-nitrogen compounds from silicic acid, carbon and nitrogen, this operation formerly had to be conducted in the electric furnace owing to the high temperature necessary. They find, however, that it is possible to conduct the reaction at a low temperature if oxides, hydroxides, or salts of metallic elements are added to the mixture of silicic acid and carbon. They also suggest the employment of silicious materials such as the silicates of iron, aluminium, calcium, etc. Instead of pure nitrogen they employ gaseous mixtures containing N.

They give as an example 75 kilos of finely divided silica mixed with 25 kilos of powdered wood charcoal, and heated in a stream of nitrogen for ten to twelve hours at 1300°-1400° C., allowing the product to cool in N gas. When the product is treated with saturated or superheated steam, ammonia is produced.

It is doubtful, however, whether any of these processes are likely to be a commercial success, in view of the successful production of synthetic ammonia directly from the elements.

However, the reader may consult the following patents :—Basset, English Patent, 4,338/97 ; Lyons and Broadwell, U.S. Patent, 816,928 ; Borchers and Beck, German Patent, 196,323 ; Roth, German Patent, 197,393 ; Kaiser, English Patent, 26,803/05 ; Wilson, English Patent, 21,755/95 ; Mehner, English Patents, 12,471/95, 2,654/97, 28,667/03.

CHAPTER VI

The Cyanamide Industry

CHAPTER VI
THE CYANAMIDE INDUSTRY

LITERATURE

SCOTT.—*Journ. Roy. Soc. Arts*, 1912, **60**, 645.

CARO.—*Chem. Trade Journ.*, 1909, **44**, 621, 641 ; *Zeit. angew. Chem.*, 1909, **22**, 1178.

FRANK.—*Zeit. angew. Chem.*, 1903, **16**, 536 ; 1905, **18**, 1734 ; 1906, **19**, 835 ; *Journ. Soc. Chem. Ind.*, 1908, **27**, 1093 ; *Trans. Faraday Soc.*, 1908, **4**, 99.

CROSSLEY.—"Thorp's Dict. of Applied Chem.," Vol. III. (1912), gives an excellent account. See also *The Pharmaceutical Journal and Pharmacist*, 12th March 1910.

Calcium Cyanamide, Nitrolime, Kalkstickstoff, CaN_2C. — Calcium carbide, CaC_2, which has been manufactured for many years for the production of acetylene gas, unites with nitrogen when heated with it to about 1000° C., forming calcium cyanamide, $CaC_2 + 2N = CaN_2C + C$.

It is supposed that a sub-carbide is first formed :—

$$CaC_2 = CaC + C ; CaC + N_2 = CaCN_2.$$

This calcium cyanamide has been found to be a valuable manure, and, containing as it does some 20 per cent. of available nitrogen against only 15 or 16 per cent. of sodium nitrate, it is now being made on an increasing scale for the manufacture of artificial fertilisers.

Works for the manufacture of the crude mixture of calcium cyanamide and carbon, known as "Nitrolime" or "Kalkstickstoff," have been erected in Norway, Germany, Italy, Austria-Hungary, Switzerland, United States, India, Japan, and other countries. The largest of these works are at Odda, on the Hardanger Fjord, Norway. The Alby United Carbide Co. combined with the Nitrogen Products & Carbide Co., Ltd., to erect new factories at Odda. These factories were supplied with 23,000 H.P. (later increased to 80,000) from the river Tysse.

Although the new industry was only begun on a commercial scale in 1907, the world's output for 1913 is estimated to be 223,600 tons, of which the works of the Nitrogen Products Co. at Odda and Alby, Sweden, are now producing together 88,000 tons per year. However, further developments are going on in Scandinavia, whereby the energy derived from three other water powers in Norway (Aura, Toke, and Blekestad-Bratland), capable of producing 600,000 H.P., of which 100,000 H.P. are now being harnessed for the annual production of 200,000 tons of calcium cyanamide. The Detti-Foss water power in Iceland—capable of generating over 400,000 H.P.—will also be utilised for the production of the same product. The Nitrogen Products and Carbide Co. claim to have at its disposal sufficient water power for an output of 2,000,000 tons of crude calcium cyanamide, which, containing as it does 20 per cent. of combined nitrogen, represents an output of fixed nitrogen equal to the whole present consumption of Chile saltpetre. These immense industrial undertakings are undoubtedly destined to change the face of the world, and entirely alter agricultural conditions of the future.

Process of Manufacture.—The calcium carbide, CaC_2, used for making the nitrolime is made by heating together in an electric furnace at about 3000° C., a mixture of lime and anthracite coal in the manner described in Martin's "Industrial Chemistry," Vol. II. The molten carbide is tapped off from the furnaces at intervals of 45 minutes, conveyed on trays to a cooling bed, then it is crushed to a powder in special grinding plant.

Calcium carbide is very hard, and the crushing is performed in several stages. In order that no risks may be run through the liberation and detonation of acetylene gas, the later stages of the crushing process are performed in an atmosphere of nitrogen. The carbide is reduced to the consistency of flour, much resembling cement in appearance.

FIG. 21.—Cyanamide Furnace House at Odda (Nitrogen Products Co.).

The finely divided carbide is then conveyed to the cyanamide furnaces for treatment with nitrogen gas.

Here the process deviates into two distinct lines of treatment, according to the type of furnace employed. We will first describe the practice at Odda, and then the practice at Westerregeln and Piano d'Orta, as typical of the two lines of treatment

At Odda the crushed carbide is placed in cylindrical vertical retorts lined with fireproof material and covered with sheet iron, and holding 300-500 kilos, as seen in our illustrations, Figs. 21 and 22

Each furnace has regulating valves and control metres. The nitrogen gas is passed into each furnace under slight pressure, and the retort is heated to about 800° by sending an electric current through carbon rods placed inside, which act as a heating resistance.

Absorption of nitrogen then takes place according to the reversible action :—

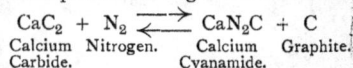

$$CaC_2 + N_2 \rightleftharpoons CaN_2C + C$$

Calcium   Nitrogen.   Calcium   Graphite.
Carbide.   Cyanamide.

The action once started proceeds with the evolution of heat, and so the temperature tends to spontaneously increase. However, the temperature of reaction must not be allowed to increase beyond 1400° C., otherwise the back action, $CaN_2C + C \longrightarrow CaC_2 + N_2$, proceeds to an increasing extent, and the cyanamide formed is thus largely destroyed again. It is essential that the temperature of reaction be kept as low as possible.

After the absorption of nitrogen has begun the heating current is switched off, as the heat developed by the action is sufficient to cause the maintenance of the proper degree of temperature. The absorption of nitrogen proceeds for thirty to forty hours, and is known to be complete by reading the controlling meter.

In this process there is no superheating, and therefore no reversal of the action. The action proceeds from the inside outwards, and thus the material shrinks inwards and away from the furnace walls. Hence the cyanamide formed is easily removed in a solid coke-like block from the furnace (after cooling in a current of air for nine hours), and is then ground to a fine powder in an air-tight grinding mill, stored in a large silo until required, and is then packed in bags with a double lining, and sent into commerce under the name " **Nitrolim.**"

The substance contains 20-22 per cent. N = 57-63 per cent. calcium cyanamide, CaN_2C, 20 per cent. lime, 14 per cent. C as graphite, 7-8 per cent. of silica, iron oxide, and alumina. The substance should be free from unchanged calcium carbide and free lime, CaO.

At Westerregeln, Piano d'Orta, and other Continental works the process of manufacture is quite different.

Here the powdered carbide is placed in horizontal retorts similar to the retorts used in making coal gas; these are heated **externally** to 800°-1000° C. by being placed in a gas-fired furnace, while a stream of nitrogen is forced into the retorts for absorption by the carbide.

Fig. 23 shows a rough diagram of the system employed in Piano d'Orta in Italy. Air is driven through retorts AA, partially filled with copper, and heated in a furnace. The oxygen of the air is absorbed by the copper, forming copper oxide ($Cu + O = CuO$), while the nitrogen passes on into the calcium carbide retort BB, heated in a furnace as shown. Here the absorption of nitrogen and resulting formation of calcium cyanamide takes place as described. The copper is regenerated by passing producer gas through the CuO.

Serious difficulties, however, have here been encountered. The absorption of nitrogen commences slowly, but then the temperature suddenly goes up as heat is developed by the action, and, unless great care in working is employed, it may exceed 1400° C., when a considerable loss of nitrogen in the formed cyanamide occurs owing to the back-action previously discussed.

Moreover, overheating of the walls ensues, and the cyanamide, on cooling, sets to a rock-like mass on the walls of the retorts, and has to be forcibly knocked out. Hence the retorts are subjected to a severe wear and tear.

These difficulties have to some extent been obviated by adding calcium chloride, $CaCl_2$ (Polzenius), or calcium fluoride, CaF_2 (Carlson), to the crushed carbide, so as to lower the temperature of reaction. With $CaCl_2$ the temperature of reaction is lowered to 700°-800°, and with CaF_2 the reaction temperature is about 900° C.

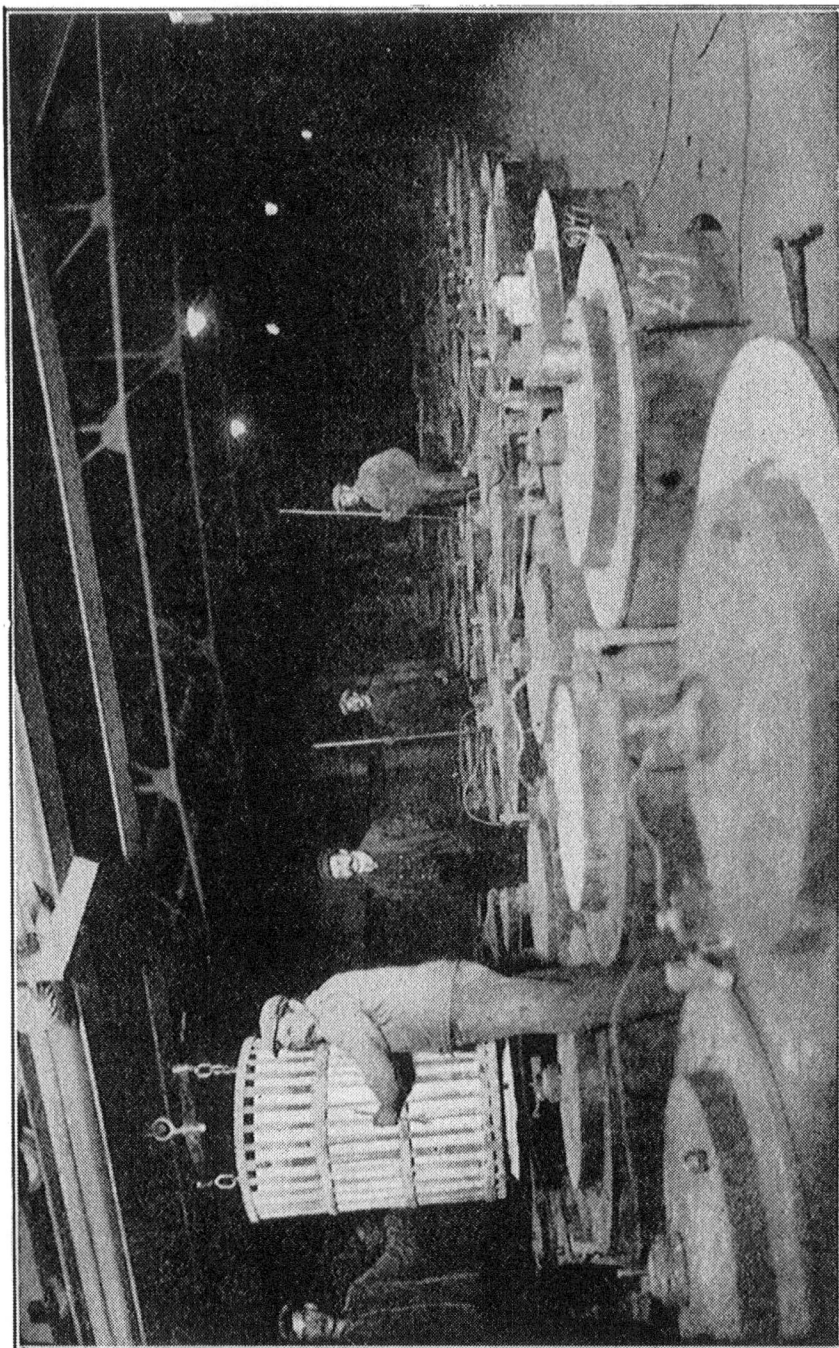

FIG. 22.—Cyanamide Furnaces at Odda, showing Carbon Resistances (Nitrogen Products Co.).

This favourable influence is supposed to be due to the fact that $CaCl_2$ or CaF_2 mixes with the cyanamide formed on the surface of the carbide, and causes it to fuse at a lower temperature than it would in a pure state. In the absence of these salts the cyanamide forms an impervious layer over the surface of the carbide, and thus prevents the entrance of the nitrogen into the interior of the carbide unless the temperature is about 1000° C. In the presence of $CaCl_2$ or CaF_2, however, the melting point of the cyanamide is much lowered; it melts or softens at 700°-800° C., and thus no longer serves as a protecting layer for hindering the absorption of the nitrogen by the interior layers of carbide.

It was supposed that this process will cause an outlet for the now practically waste $CaCl_2$ produced as a by-product in many chemical operations.

The Gesellschaft für Stickstoffdünger, at Westerregeln, add anything up to 10 per cent. $CaCl_2$ to the powdered carbide, the absorption temperature of the N being only 700° C. The product, containing 20 per cent. N, 45 per cent. Ca, 19.5 per cent. C, 6.5 per cent. Cl, and 9 per cent. impurities, is put on the market under the name "Stickstoffkalk."

The disadvantage of the presence of $CaCl_2$ is that the product is hygroscopic, and so cannot be stored easily. It is, therefore, sold principally in spring, but it is stated that the process has been abandoned (1910).

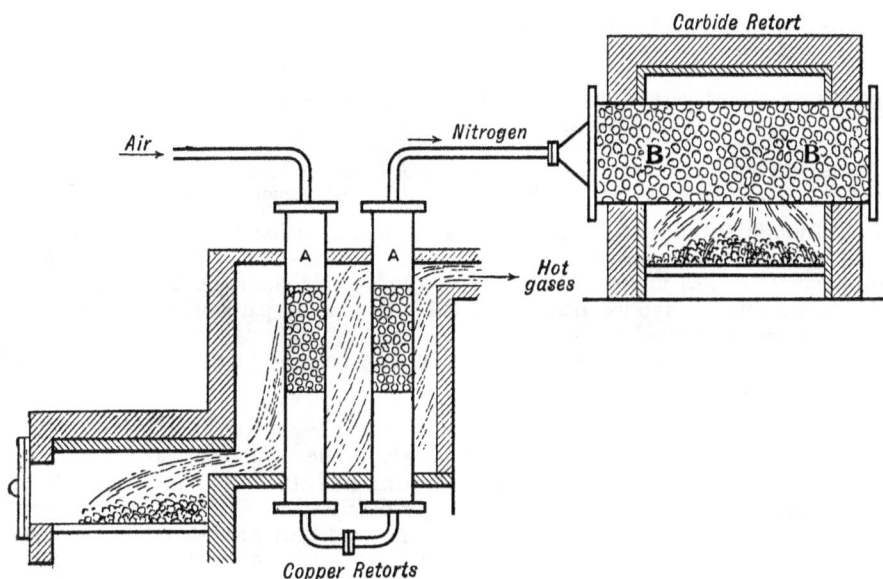

FIG. 22A.

The **Nitrogen** required for the manufacture of nitrolime is obtained at Odda by liquefying the atmosphere, and separating the oxygen and nitrogen by fractional distillation.

The Linde plant used at Odda for the purpose is the largest in the world, liquefying 100 tons of air daily, from which about 77 tons of nitrogen are obtained.

The new works at Odda, however, are fitted up with a Claude plant, which is stated to be simpler to work than the Linde.

In some Continental works the nitrogen is produced by passing air over hot copper, which retains the oxygen as copper oxide, and allows the nitrogen to pass on, as explained above. The copper is regenerated by passing water gas over the treated copper.

The nitrogen produced must be free from moisture and oxygen. The Linde plant already in use gives 100,000 cub. ft. of nitrogen per hour, containing less than 0.4 per cent. oxygen. In the Claude machines, now installed in the new works at Odda, the nitrogen supplied contains less than 0.1 per cent. O. The presence of moisture would decompose the calcium carbide thus:—

$$CaC_2 + 2H_2O = Ca(OH)_2 + C_2H_2.$$

5

Oxygen would decompose the calcium cyanamide, forming $CaCO_3$, thus reducing the percentage of nitrogen in the finished product. CO_2 or CO, if present in the air, would act thus on the carbide :—

$$2CaC_2 + CO_2 = 2CaO + 5C. \qquad CaC_2 + CO = CaO + 3C.$$

Uses of Calcium Cyanamide.—This body is mainly used as a **nitrogenous manure**. When placed in the soil under suitable conditions its nitrogen is set free in the form of ammonia by means of bacteria. The end results can be expressed by the equation :—

$$CaNCN + 3H_2O = 2NH_3 + CaCO_3,$$

and so it may be used directly as a substitute for ammonium sulphate or Chile saltpetre.

The decomposition of the cyanamide in the soil undoubtedly takes place in stages. Atmospheric CO_2 and moisture first liberate cyanamide, thus :—

$$CaNCN + H_2O + CO_2 = CaCO_3 + H_2NCN.$$

The cyanamide is then converted into urea :—

$$H_2NCN + H_2O = CO(HN_2)_2,$$

which is speedily transformed by bacterial fermentation into ammonium carbonate :—

$$CO(NH_2)_2 + 2H_2O = (NH_4)_2CO_3.$$

This latter is either directly absorbed by the plants, or is first oxidised to nitrate and nitrite by nitrifying organisms in the manner explained in Chapter I. ("Circulation of Nitrogen").

Calcium cyanamide must be applied as a manure below the surface of the soil some time before the seed is sown. It must not be applied to plants actually growing. It is also mixed with Bessemer slag or potassium salts as a component of mixed manures. 150 lbs. per acre, mixed with 100-125 parts of potassium salts, is the mixture most used.

Calcium cyanamide, however, also serves as the basis for the manufacture of many nitrogenous compounds. For example :—

Ammonia, NH_3, is manufactured from it by treating it with superheated steam ($CaNCN + 3H_2O = CaCO_3 + 2NH_3$), see p. 56. The ammonia can be easily combined with nitric acid to form **ammonium nitrate**. **Nitric acid** itself is now manufactured from the ammonia by oxidising it by **Ostwald's Process** (p. 28). From the ammonium nitrite and nitric acid thus produced, **explosives**, **dyes**, and other useful nitrogenous substances are now made. The manufacture of **cyanides** from calcium cyanamide is also now an important industry (see p. 78). Sodium cyanide, NaCN, is now produced by fusing calcium cyanamide with common salt, 90-95 per cent. of the cyanamide being converted into sodium cyanide, which is much used for recovery of gold in South Africa, Australia, U.S.A., Mexico, etc. Calcium cyanamide is stated by Clancy (*Metall. Chem. Eng.*, 1910, **8**, 608, 623; 1911, **9**, 21, 53, 123) to be capable of replacing cyanides in gold extraction.

Dicyandiamide is made in a crystalline form by extracting technical calcium cyanamide with hot water :—

$$2CaNCN + 4H_2O = 2Ca(OH)_2 + (H_2N.CN)_2.$$

It is used for making organic dyes, also for reducing the temperature of combustion of explosives, since its decomposition produces little heat, and gives a strong pressure owing to it containing 66.6 inert N. Hence the addition of the substance to powers like cordite, which rapidly destroy rifling on account of high temperature of combustion (see Chapter IX., "Explosives").

Urea, $CO(NH_2)_2$, **Guanidine**, $C(NH)(NH_2)_2$, **nitroguanidine**, etc., are also nitrogenous substances manufactured at Spandau by treating calcium cyanamide with aqueous acids, etc.

The body, "**ferrodur**," used for case hardening and tempering iron and steel, contains nitrolim.

The whole family of nitrogen products may be traced in the following table :—

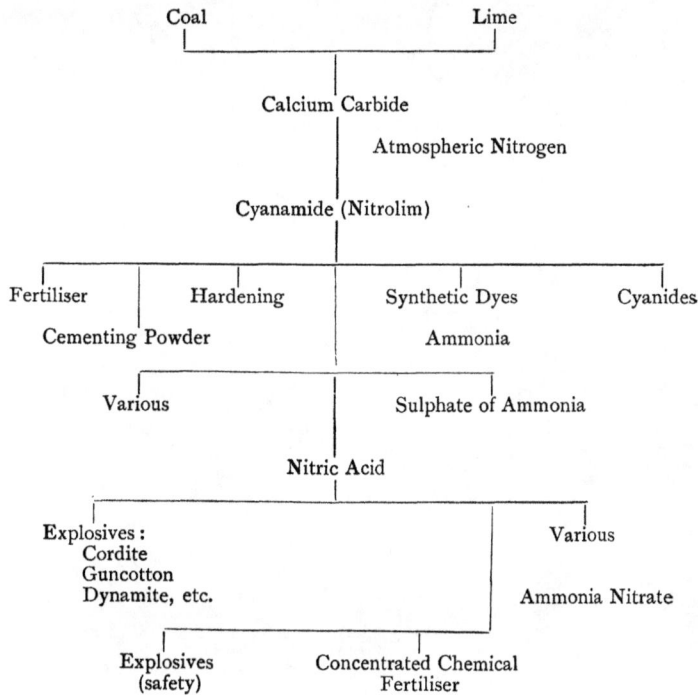

```
            Coal                        Lime
             |_____|
                          |
                   Calcium Carbide

                          Atmospheric Nitrogen
                          |
                 Cyanamide (Nitrolim)
                          |
     |_____|_____|_____|
     |           |            |              |
 Fertiliser  Hardening   Synthetic Dyes    Cyanides
        Cementing Powder        Ammonia
              |                    |
              |_____|
              |                    |
          Various          Sulphate of Ammonia
                          |
                     Nitric Acid
              |_____|
              |                         |
      Explosives :                   Various
         Cordite
         Guncotton
         Dynamite, etc.          Ammonia Nitrate
              |_____|
              |                         |
         Explosives          Concentrated Chemical
         (safety)                  Fertiliser
```

CHAPTER VII

———

The Cyanide and Prussiate Industry

CHAPTER VII

THE CYANIDE AND PRUSSIATE INDUSTRY

LITERATURE

BERTELSMANN.—" Technologie der Cyanverbindungen " (1906).

LUNGE.—"Coal Tar and Ammonia." 4th Edition.

T. EWAN.—"Cyanides" in Thorpe's "Dict. Appl. Chem.," vol. ii., 1913.

BEILBY.—*Journ. Soc. Chem. Ind.*, 1898, 28th Feb.

CONROY.—*Journ. Soc. Chem. Ind.*, 1899, **18**, 432.

OST.—*Zeit. angew. Chem.*, 1906, **19**, 609.

G. ERLWEIN.—" Fifth Internat. Kongress angew. Chem." Berlin, 1903, I., 646.

P. E. WILLIAMS.—*Journ. Gas-Lighting*, 1st Oct. 1912, p. 31 *et seq.*

HOFMANN and CO-WORKERS.—*Annalen*, 1904, **337**, 1 ; 1905, **340**, 267 ; 1905, **342**, 364.

Also references and patents in the text.

Until the year 1890 only comparatively small quantities of the very poisonous potassium cyanide, KCN, were made, being used principally in the electroplating industry and in photography.

This use was founded on the fact that KCN possesses the property of dissolving the salts of metals like gold, silver, and nickel, etc., the metal going into solution as a double cyanide. From these solutions the metal can be readily deposited electro-chemically in a coherent layer by making the surface of the object to be coated with the precious metal the negative pole in a solution of the double cyanide.

Up to 1890 the annual output of KCN was about 100 tons, the product being usually prepared by fusing ferrocyanides with alkali.

In 1887 M'Arthur and Forrest took out their patents (English Patents, 14,174/87 ; 10223/88) for using dilute KCN or NaCN solutions for extracting gold from ores, and in a few years from that date the manufacture of cyanides on an industrial scale was well established.

In 1903 about 6,300 tons of cyanide were placed on the market.

In 1910 Great Britain exported of cyanide of sodium or potassium, 7,770 tons, value £633,000. Germany, in 1895, exported 6,280 tons at 1,400 marks per ton.

In 1895 only 120 tons at 3,200 marks were produced. The German production of ferrocyanide of potassium of sodium was in 1895 valued at 361 tons at 1,480 marks. In 1909 this had become 1,450 tons at 870 marks per ton.

The development of the vast Transvaal gold industry has been mainly dependent upon the discovery that gold can be dissolved out of its ores by treating with dilute cyanide solution. The peculiar formation of " blanket " ore, the small proportion of gold per ton which it contains, the finely divided state of the gold in these ores, as well as the production of " slimes " and " tailings " containing some gold, all combined to render the ordinary methods of gold extraction difficult of application.

When these gold-bearing materials are treated with dilute KCN solution the gold passes into solution, thus :—

$$4KCN + 2Au + H_2O + O = 2KAu(CN)_2 + KHO$$

The gold thus passes into solution as potassium aurocyanide, $KAu(CN)_2$.

The gold is next precipitated on zinc according to the equation :—

$$2KAu(CN)_2 + Zn = K_2Zn(CN)_2 + 2Au.$$

The maximum effective solution is that containing 0.25 per cent. KCN, solutions containing as little as 0·0005 per cent. KCN having a perceptible solvent action. The quantity of solution used is 33 per cent. of the amount of ore in the charge.

Previous to the introduction of the process into South Africa no tailings had been treated by lixiviation with the KCN solution. The "Robinson" works put up a plant costing £3,000, and made £2,000 **per month** working profit !

This great success led to the extension of the process to all the gold-bearing countries of the world, such as Australia, New Zealand, India, America, etc. In 1898 the cyanide treatment of rand "tailings" cost 2s. 3d. to 2s. 9d. per ton, using ¼lb. to ½lb. of cyanide per ton ; this cyanide cost 3d. to 6d. at that date. Since then the cost of the cyanide has considerably decreased. According to Beilby (*Journ. of the Soc. of Chem. Ind.*, 28th Feb. 1898), the total production of gold by the cyanide process in 1897 was 1,215,000 oz. of fine gold. In 1910 the world's output of gold was 23,000,000 oz. of fine gold, 5,750,000 oz. of this (about 25 per cent.) being recovered by the use of cyanide. The value of the cyanide process may be judged from the fact that the value of the cyanide bullion produced by the rand in 1910 was equal to the whole profit earned by the mines. The actual figures being :—

Yield from cyaniding	-	-	-	£11,552,743
Total profit of working	-	-	-	£11,567,099

MANUFACTURE OF SODIUM OR POTASSIUM CYANIDE

Potassium cyanide or sodium cyanide is now manufactured by several different sources, viz. :—

1. From ferrocyanide or ferricyanide (prussiate).
2. From ammonia, carbon, and an alkali metal (Na), or alkali salt.
3. From sulphocyanides.
4. From "schlempe," a residue from the refining of beet sugar.
5. From cyanamide (nitrolim).
6. From atmospheric nitrogen.

MANUFACTURE OF CYANIDE FROM FERROCYANIDES (PRUSSIATE)

Previous to the introduction of the gold-cyaniding process almost all the cyanide manufactured was made from sodium or potassium ferrocyanide, which was produced from animal matter, such as dried blood, horns, hoofs, and the residues of slaughter houses (see under **Ferrocyanides**, p. 79).

When potassium ferrocyanide thus obtained is heated with well-dried potassium carbonate, the following action occurs :—

$$K_4Fe(CN)_6 + K_2CO_3 = \underset{\substack{\text{Potassium}\\\text{cyanide.}}}{5KCN} + \underset{\substack{\text{Potassium}\\\text{cyanate.}}}{KCNO} + CO_2 + Fe.$$

When this process was employed, potassium cyanide sold at 2s. 6d. per lb. as against 7½d. per lb. in 1911. Only about 20 per cent. of the nitrogen of the animal matter was used, and barely 80 per cent. of the potash, the remainder being lost.

Potassium (or sodium) ferrocyanide is no longer made from refuse animal matter, but is now recovered from gasworks ; the recovered ferrocyanide ("prussiate") is largely worked for the manufacture of cyanide by **Erlenmeyer's process of fusing with metallic sodium,** when the following changes take place :—

$$K_4Fe(CN)_6 + 2Na = 4KCN + 2NaCN + Fe.$$

In this process, first worked between 1890-1900, all the cyanogen is recovered in the form of sodium or potassium cyanide, the sodium cyanide being technically of the same value as the potassium cyanide, provided the CN content is the same.

The process is carried out as follows :—In covered iron crucibles, some 30-40 cm. in height, dehydrated potassium ferrocyanide, $K_4Fe(CN)_6$, is mixed with the proper amount of metallic sodium in the form of short bars, and the crucible is then heated over a free fire until the contents are completely fused. The molten contents of a number of these crucibles are next poured into an iron crucible, heated by direct fire as before, but provided with a filtering arrangement made of spongy iron (obtained in the above-mentioned melting process), below which are outlet tubes. The molten cyanide is forced through this filter by means of compressed air and a compressing piston. The filtered cyanide as it flows away from the filtering crucible solidifies to a white crystalline mass. It contains some cyanate, KCNO or NaCNO, along with a little alkali carbonate and free caustic soda or potash. Nevertheless, in practice the cyanide is always valued on the basis of the KCN it contains, and since 75.3 parts of NaCN are technically equivalent to 100 g. KCN, the cyanide can be placed on the market as " 100 per cent. KCN" in spite of the presence of these impurities. It is only the CN which counts, technically ; whether the CN is united with K or Na is immaterial.

It was this process of cyanide manufacture which created the sodium industry, the sodium used being principally produced by the Castner electrolytic process The production of metallic sodium is now an integral part of the cyanide industry. It is described in Vol. III. of the work devoted to metallurgy.

The use of sodium resulted in the production of a sodium-potassium cyanide, the ratio of sodium to potassium being varied within wide limits. The larger the amount of Na salt present the higher the strength of the cyanide. At first considerable opposition was encountered to the use of a cyanide containing sodium, owing to commercial reasons. It has been shown, however, that sodium cyanide is as effective as potassium cyanide, and is cheaper ; consequently, whereas formerly only potassium ferrocyanide, $K_4Fe(CN)_6$, was manufactured, at the present time it has been almost entirely superseded (since 1905) by the cheaper sodium ferrocyanide (prussiate), $Na_4Fe(CN)_6$, which can be obtained very pure, and yields, when fused with metallic sodium as above described, practically pure sodium cyanide, NaCN.

Commercial sodium cyanide, NaCN, containing as it does about 30 per cent. more cyanogen than KCN, can be used in smaller quantities than KCN for producing the same gold-dissolving effect. Moreover, it is stated to be more convenient for making up solutions.

MANUFACTURE OF CYANIDES FROM AMMONIA, CARBON, AND ALKALI METAL, OR ALKALI SALT

Several important processes are worked.

Siepermann's Process (see English Patents 13,697 of 1889, 9,350 and 9,351 of 1900). One part of sodium carbonate and two parts of charcoal (that is, sufficient charcoal to keep the mass from fusing during the process) is heated to dark redness in the upper part of a vertical iron tube while a current of ammonia gas is sent through the mixture. Potassium cyanate, KCNO, is formed thus :—

$$K_2CO_3 + NH_3 = KCNO + KOH + H_2O.$$

The mixture is then allowed to fall to the bottom of the tube, where it is heated to bright redness. The cyanate is decomposed with formation of cyanide :—

$$KCNO + C = KCN + CO.$$

The final product is thrown into air-tight vessels, cooled, lixiviated with water, the solution being evaporated *in vacuo* until the KCN crystallises out. KCN is soluble with difficulty in the presence of much KOH or K_2CO_3, and crystallises out before these salts in the form of anhydrous crystals. As first made the KCN was a damp deliquescent mass, which had to be fused with the product of the ferrocyanide process. The working of the process is difficult. It has been worked at Stassfurt since 1892.

Bielby Process (see English Patent, 4,820 of 1091).—The principle is much the same as the **Siepermann Process**, but differs in important details.

Much less carbon is used, so that at the end only slight excess remains. The charcoal is added gradually during the operation, so that the material is always present as a molten liquid through which the ammonia gas is forced under a slight pressure, when the following action takes place :—

$$K_2CO_3 + 4C + 2NH_3 = 2KCN + 3CO + 3H_2.$$

The final molten product is filtered from the small excess of unchanged charcoal, and thus a white saleable product is directly obtained without the difficulties of lixiviation. However, since the melting point of the pure potassium carbonate is inconveniently high (about 890° C.), ready-made cyanide is added to it in order to reduce the temperature of fusion.

The Beilby process has been worked since 1892 by the Cassel Gold Extracting Co. at Glasgow, and has achieved remarkable success.

In 1899 Beilby's process was estimated to supply fully 50 per cent. of the world's output of high-strength cyanide.

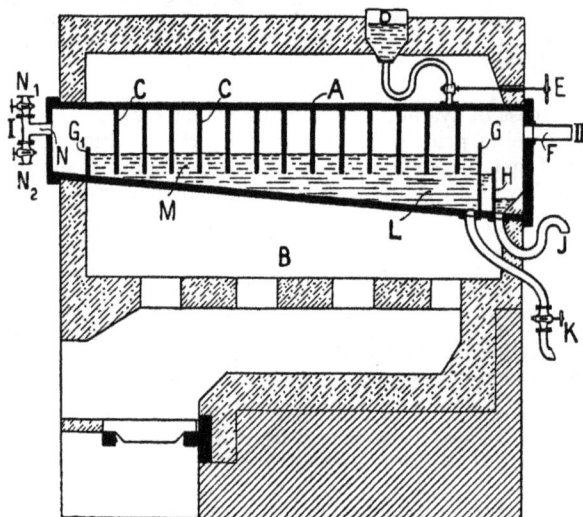

FIG. 23.—Castner's Sodamide Furnace (Longitudinal Section).

The Castner Process (see English Patents, 12,219 of 1894; 21,732 of 1894. See also the German Patents, 117,623, 124,977 (1900); 126,241 (1900); 148,045 (1901).—One of the most important syntheses of cyanide from ammonia was successfully worked on the large scale by the **Frankfurter Scheideanstalt** in 1900, the process having been worked out by H. Y. Castner some years

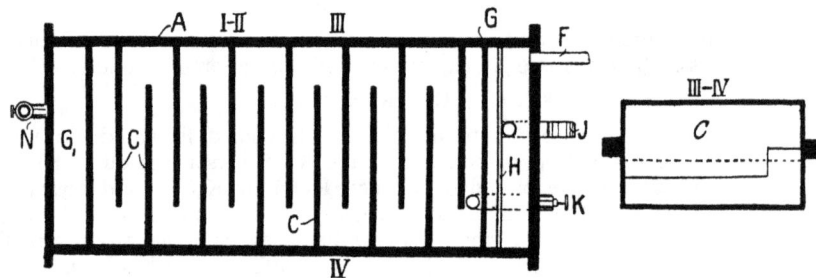

FIG. 24.—Castner's Sodamide Furnace (Horizontal Section).

previously. Dry ammonia gas led over molten sodium in the absence of air yields sodamide, $NaNH_2$, as a crystalline product, thus :—

$$NH_3 + Na = NaNH_2 + H.$$

On adding to the fused sodamide some powdered charcoal or coal, sodium cyanamide, Na_2N_2C, is formed, thus :—

$$2NaNH_2 + C = Na_2N_2C + 4H.$$

At a still higher temperature the excess of carbon reacts with the cyanamide to form cyanide, thus :—

$$Na_2N_2C + C = 2NaCN.$$

Acetylene gas (German Patent, 149,678, 1901) can be used instead of carbon as the reducing agent. Like the Beilby process, the final product is a fused mass containing very small quantities of solid impurities, which are easily removed by filtration. The operations involved are simple, but the temperatures employed at each stage of manufacture must be carefully regulated. The process is carried out as follows :—

Into an iron crucible heated to 500° C. some 70 kg. of wood charcoal is discharged, and the whole is heated in a slow stream of ammonia gas. Next 115 kg. of metallic sodium is added to the crucible, the current of ammonia is increased, and the temperature is raised to 600° C. until all the sodium is converted into sodamide. The temperature is finally increased to 800° C., when the carbon acts on the sodamide and converts it into cyanide, with intermediate formation of sodium cyanamide, as above explained. The process proceeds quantitatively, and the final molten cyanide is poured off and filtered while still in a fluid condition, and cast in iron moulds. It then forms pure white cakes, containing 97.5-98 per cent. NaCN (equivalent to 128-130 per cent. KCN).

Figs. 24 and 25 show Castner's sodamide furnace. It consists of an iron retort A, the upper part of which is provided with a series of vertical baffle plates C, arranged as indicated. The retorts are heated to 300°-400° C., ammonia entering at N and molten metallic sodium at D, while the fused fuel products can be run off by K.

This process, going hand in hand with the fall in the price of metallic sodium, caused an enormous reduction in the price of cyanide from £160 per ton in 1895 to only £70 per ton in 1909. No doubt, in consequence of the production of cheap synthetic ammonia (see page 53), a further fall in price of cyanide may be anticipated.

MANUFACTURE OF CYANIDE FROM SULPHOCYANIDES

Many attempts to produce cyanides directly from sulphocyanides without the intermediate production of ferrocyanides have been proposed, and some have been worked on the industrial scale, without, however, producing more than a small amount of the total cyanide.

Playfair, in 1890 (English Patent, 7,764 of 1890, see also *Journ. Soc. Chem. Ind.*, **11**, 14 (1892); Conroy, *Ibid.*, **15**, 8 (1896), found that fusing at 400° C. lead or zinc with sulphocyanide abstracted the sulphur, leaving cyanide $NaCNS + Pb = NaCN + PbS$.

Ferrocyanide (prussiate) was produced from sulphocyanide by treating with finely divided iron ("swarf") :—

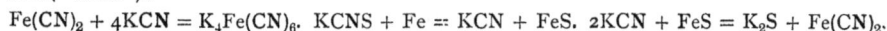

$Fe(CN)_2 + 4KCN = K_4Fe(CN)_6$. $KCNS + Fe = KCN + FeS$. $2KCN + FeS = K_2S + Fe(CN)_2$.

The usual process was to heat the sulphocyanide in retorts to a dull red heat, cool in absence of air, and extract the product with water in the retorts themselves, so that contact with air, and subsequent oxidation, are thus entirely avoided.

This process, or some modification, has repeatedly been worked since 1860, when Gelis took out his patent (English Patent, 1,816 of 1860). See also the following patents :—English Patents, 1,148/78, 1,359/79, 1,261/81, 5,830/94. The British Cyanide Co. have worked the process at Oldbury since 1894.

A different method of work is suggested in the patents 361/96, where copper is used instead of iron. In the German Patent, 32,892 of 1882, KCNS is heated with iron filings, ferrous hydroxide, and water to 110°-120° :—

$$6KCNS + 6Fe + Fe(OH)_2 = K_4Fe(CN)_6 + 6FeS + 2KOH.$$

Raschen, in 1895 (see English Patents, 10,476/95, 10,956/95, 21,678/95, 19,767/98, 12,180/1900 ; see also Conroy, *Journ. Soc. Chem. Ind.*, 1899, **18**, 432), invented a daring method of converting sulphocyanide into cyanide, which was worked by the United Alkali Co. A 15 per cent. NaCNS solution flows into boiling dilute HNO_3, when HCN and NO is formed thus :—

$$NaCNS + 2HNO_3 = HCN + 2NO + NaHSO_4.$$

The hydrocyanic acid gas, together with the nitric oxide, is passed through alkali, whereby all the HCN is absorbed as alkaline cyanide, and is recovered by evaporating. The nitric oxide passes forward, mixed with air, through towers packed with flints down which water trickles. It is thus converted into nitric acid once more, and is thus used again.

The corrosive natures of the solutions employed, combined with the enormous volumes of highly poisonous gas to be dealt with, presented engineering problems of great difficulty which, however, were successfully overcome. The sulphocyanide was produced synthetically from carbon bisulphide and ammonia.

It should be noted that sulphocyanides can be directly reduced to cyanides by treating with hydrogen at a dull red heat:—

$$KSCN + H_2 = KCN + H_2S. \qquad 2KCNS + 2H_2 = K_2S + 2HCN + H_2S.$$

Seventy per cent. of the N is obtained as cyanide, and 20 per cent. as HCN gas. See Playfair, *Journ. Soc. Chem. Ind.*, 1892, **11**, 14; Conroy, Heslop, and Shores, *loc. cit.*, 1901, **20**, 320.

MANUFACTURE OF CYANIDES FROM "SCHLEMPE"

Bueb of Dessau has introduced a successful process for extracting cyanide by heating "schlempe," a waste product of beetroot sugar manufacture ("Industrial Chemistry," Vol. I.).

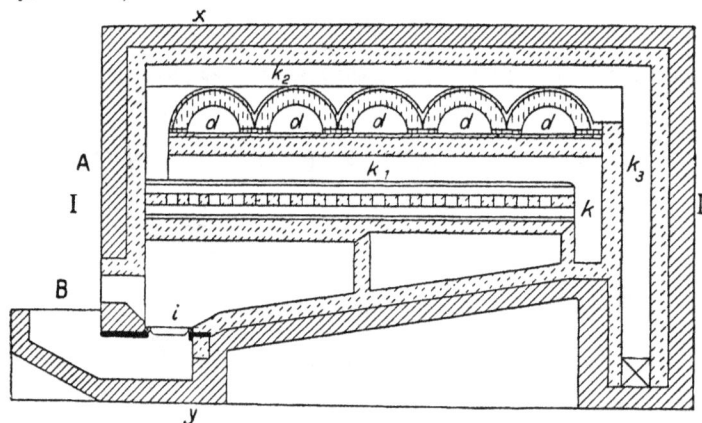

FIG. 26.

The beet juice is clarified, and as much as possible of the crystalline sugar is separated by evaporation in vacuum pans and crystallisation. There remains a thick, viscid, highly coloured, and strongly smelling liquid, still containing 50 per cent. sugar, 20 per cent. organic impurities, 10 per cent. salts, and 20 per cent. water. This is beet sugar molasses. To extract more sugar from it the molasses is treated with alkaline earths, when the sugar combines with the base to form feeble salts, known as "sucrates," which crystallise in well-defined crystals, and are easily separated from the adhering molasses. The sugar is set free from the sucrates by treating with carbon dioxide gas in excess of water, or simple excess of water, the alkaline earth separating from the sugar as an insoluble precipitate. The brown liquid from which the sugar has been separated contains all the impurities of the original beet juice, together with much potash and other salts. It is called "schlempe," and contains about 4 per cent. of nitrogen.

A few years ago this schlempe was simply burnt in furnaces so as to obtain the potash salts in the form of a residue known as "schlempe kohle," all the nitrogen being lost.

Vincent in France, about 1878, attempted to utilise and recover the nitrogen at a red heat, when ammonia, methylamines, methyl alcohols, and gases were obtained The manufacture of the methylamines was abandoned in 1881 for want of a market.

In 1898, however, Bueb (see English Patent, 7,171/95; 26,259/98) introduced his process of working the schlempe for cyanide and ammonia in a molasses refinery at Dessau, and the process is now worked in a great many factories.

The schlempe, concentrated to 1.4 sp. gr., is run into earthenware retorts, and subjected to dry destructive distillation.

The gases which escape, consisting of CO_2, CO, H, CH_4, NH_3, N_2, and much trimethylamine, $N(CH_3)_3$ (the latter being derived from betaine, $HO.N(CH_3)_3 CH_2COOH$, contained in the schlempe), do not at this stage contain any cyanide. However, when these gases are passed into "super-heaters," consisting of cylindrical chambers filled with brick chequer-work maintained at a red heat (about 1000° C.), hydrocyanic acid is obtained by direct decomposition of the methylamine, thus :—

$$N(CH_3)_3 = HCN + 2CH_4.$$

The superheating of the gases must not last long, as HCN is unstable and the yield would soon diminish. As a rule, two chambers are used alternately, one being heated while the gases from the retort are passing through the other.

Figs. 25, 26, 27 show Bueb's plant. *d, d, d* are the schlempe retorts, which are heated to 700°-800° C. by the furnace gases passing through the flues $K_1K_2.K$. The superheater lies beneath the retorts as indicated, and is heated to about 1,000° C by the furnace gases. The charge for each retort consists of 180 kg. of schlempe, and the distillation takes three to four hours.

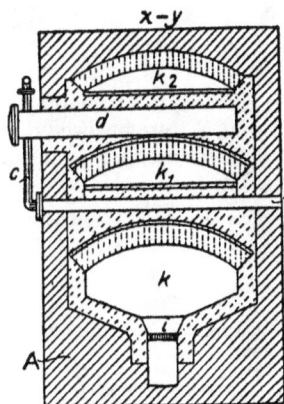

FIG. 27.

The "cyanised" gas leaving the decomposition chambers, containing 10 per cent. HCN and 5-8 per cent. NH_3 by volume, passes through a series of coolers, and then into dilute sulphuric acid, where the ammonia, pyridine, and similar bases are absorbed.

Next they pass through a "cyanide absorber," where the HCN is absorbed in water, and, after treating with NaOH, is obtained as a concentrated solution of NaCN.

The combustible gases escaping this treatment are burnt under the retorts, and so are used for heating purposes. All the retorts and leading tubes are worked under a somewhat diminished pressure, so that no poisonous gases escape into the air. The issuing gas is tested for traces of HCN by passing through a NaOH solution to which a little $FeSO_4$ has been added, when the slightest trace of HCN is revealed by a precipitate of Prussian blue.

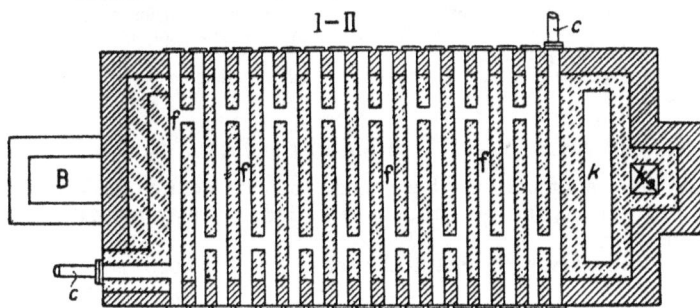

FIG. 28.

The concentrated aqueous solution of NaCN, obtained as above described, as a weakly alkaline solution, is evaporated in vacuum vessels, and allowed to crystallise in a nodular form, a temperature above 30° C. being essential, as under this temperature the NaCN separates as $NaCN.2H_2O$. The crystals are centrifuged, dried, and pressed to hard cakes, containing sodium cyanide equivalent to 120-125 per cent. KCN.

About 35 per cent. of the N in the schlempe is thus recovered as cyanide, 25 per cent. as NH_3, while 40 per cent. is lost in the form of nitrogen gas.

Ost (see *Zeit. angew. Chem.*, 1906, **19**, 609), calculates that from the 13,000,000 tons of beetroot annually worked in Germany for sugar, about 10,000 tons of KCN would be obtainable from the schlempe by Bueb's process. Otherwise the N in schlempe has no commercial value.

MANUFACTURE OF CYANIDES FROM CALCIUM CYANAMIDE ($CaCN_2$)

When the mixture of calcium cyanamide and carbon ($CaCN_2 + C$), which is known commercially as **nitrolim** or **kalkstickstoff** (p. 61), is fused either with salt, NaCl, or sodium carbonate, Na_2CO_3, it is converted to the extent of 90-95 per cent. into sodium cyanide, NaCN, thus :—

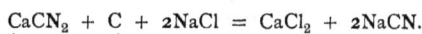

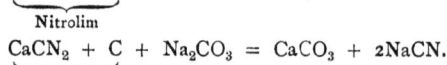

$$\underbrace{CaCN_2 + C}_{\text{Nitrolim}} + 2NaCl = CaCl_2 + 2NaCN.$$

$$\underbrace{CaCN_2 + C}_{} + Na_2CO_3 = CaCO_3 + 2NaCN.$$

In practice, salt is employed for fusion; the product, containing about 30 per cent. of NaCN, can be used directly for gold extraction. When pure NaCN is required the fused mixture is decomposed by acids and the liberated HCN absorbed in NaOH.

The formation of cyanide from cyanamide is reversible :—

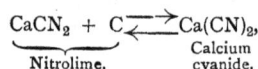

$$\underbrace{CaCN_2 + C}_{\text{Nitrolime.}} \underset{\longleftarrow}{\overset{\longrightarrow}{}} \underset{\substack{\text{Calcium}\\\text{cyanide.}}}{Ca(CN)_2,}$$

so that special precautions have to be adopted to prevent reformation of cyanamide after complete fusion.

Moreover, at the temperature of fusion the cyanamide may be partially reconverted into calcium carbide with loss of nitrogen by the reaction :—

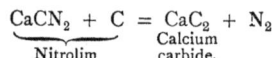

$$\underbrace{CaCN_2 + C}_{\text{Nitrolim}} = \underset{\substack{\text{Calcium}\\\text{carbide.}}}{CaC_2} + N_2$$

The liberated carbide then destroys some cyanide by reconverting it into cyanamide, thus :—

$$Ca(CN)_2 + CaC_2 = 2CaCN_2 + 2C.$$

All these side reactions tend to diminish the yield of cyanide. Technically these difficulties are stated to have been overcome by "using appropriate appliances for melting and cooling the materials" (although no details have been published), and the conversion of cyanamide into cyanide is stated to be practically quantitative.

The Nitrogen Products Co. state that they use metallic sodium in the manufacture of cyanide from cyanamide.

CYANIDES FROM ATMOSPHERIC NITROGEN

A great many attempts have been made to manufacture cyanides directly from atmospheric nitrogen, and no doubt a large industry may ultimately result from these attempts.

Up to 1913, however, no great technical success has been recorded.

Possoz and Boissière's process (English Patent, 9,985/1843) consists in soaking wood charcoal in KOH, drying and heating to bright redness in retorts through which a mixture of nitrogen and CO_2 and CO gas (furnace gas) was passed. Nitrogen was taken up, 50 per cent. of the alkali in the charcoal being converted into cyanide, which was finally extracted by lixiviation. The process was not a commercial success.

Later, in 1860, Marqueritte and De Sourdeval (English Patent, 1,171/1860) attempted to use barium oxide and carbon for absorbing nitrogen, and Mond (English Patent, 433/1882) worked up a process in detail, $Ba(CN)_2$ being formed best at a temperature of 1400° C. Readman (English Patent, 6,621/94), using the same materials, but with electrical heating, worked the process at the Scottish Cyanide Co. works between 1899-1907, but although much cyanide was produced, the venture was not commercially successful.

Frank and Caro (English Patent, 15,066/95) and Wilson (English Patent, 21,997/95) found that the carbides of the alkaline earths absorb nitrogen, and attempts have been made to manufacture cyanides from the product.

Finely divided barium carbide, heated at 700° C. in nitrogen gas, takes up 11 per cent. of N, 30 per cent. of which is fixed as barium cyanide, $Ba(CN)_2$, the rest being present as barium cyanamide, BaN_2C. The material is heated with dry Na_2CO_3 and C. when the cyanamide is converted into cyanide, the change being, according to Dreschel :—

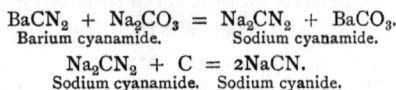

$$BaCN_2 + Na_2CO_3 = Na_2CN_2 + BaCO_3.$$
Barium cyanamide.   Sodium cyanamide.

$$Na_2CN_2 + C = 2NaCN.$$
Sodium cyanamide. Sodium cyanide.

The product is extracted with water, the cyanide is converted into ferrocyanide, fused with sodium, and 86 per cent. of the fixed nitrogen is thus converted into saleable sodium cyanide. The process was not commercially successful.

FERROCYANIDES (PRUSSIATES)

Potassium Ferrocyanide, $K_4Fe(CN)_6 + 3H_2O$, and **Sodium Ferrocyanide**, $Na_4Fe(CN)_6 + 12H_2O$, both crystallise in large yellow crystals, are moderately soluble in cold water, very soluble in hot, and possess the valuable property of not being poisonous. They are completely dehydrated above 100° C., and decompose at red heat, forming KCN, C, and Fe, N escaping.

Solubility of $K_4Fe(CN)_6$.—100 parts water dissolve (Étard, *Ann. Chem. Phys.*, 1894 [7], **2**, 546) :—

Temperature°	0°	20°	40°	60°	75°	80°	89°	98°	157° C.
Grains $K_4Fe(CN)_6$	14.5	24.5	36	49.5	64	70	72	74	88.

Solubility of $Na_4Fe(CN)_6$.—100 parts water dissolve (Conroy, *Journ. Soc. Chem. Ind.*, 1898, **17**, 103) :—

Temperature°	20°	30°	40°	50°	60°	70°	80°	90°	100° C.
Grams $K_4Fe(CN)_6$	17.9	23.5	29	35.5	42.5	51.5	59.2	61	63.

Ferrocyanides are easily produced when solutions of potassium or sodium cyanides are brought into contact with ferrous hydrate or ferrous sulphide :—

$$6KCN + FeS = K_4Fe(CN)_6 + K_2S.$$

Old Process of Manufacture.—This is practically obsolete, although a little ferrocyanide is still produced therefrom.

Into iron crucibles containing molten potash, scraps of nitrogenous animal matter, such as horns, hoofs, hair, wool, etc., are thrown. KCN is thus formed. Into the mass iron fillings are introduced, which at once unite with the organic sulphur present to form FeS. This iron sulphide then unites with the KCN to form ferrocyanide, as explained above.

The mass is extracted with water, the ferrocyanide going into solution, and being finally recovered by crystallisation. The mother liquors are used over again.

Only a poor yield is thus obtained. The residue left after lixiviation is a black mass, possessing powerful decolorising properties; it is used for decolorising paraffin and ceresin wax. In fact, the manufacture of the "animal black charcoal" is at present the main object of the industry.

Modern Process for Ferrocyanide Manufacture from Coal Gas.—Practically all the ferrocyanide of to-day is recovered from **coal gas**. 100 kg. of coal yield on destructive distillation some 30-40 g. of hydrocyanic acid gas, HCN. In 100 volumes of unpurified coal gas there are 0.1-0.2 per cent. by volume of HCN gas. In general, of the 1-2 per cent. N found in coal, some 15 per cent. is converted into NH_3, $2\frac{1}{2}$ per cent. as HCN, and 48 per cent. remains behind in the coke. The rest of the N escapes as such. The exact amount of HCN produced depends upon many factors, such as the moisture in the coal (which acts unfavourably), the rapidity and length of heating, etc. Rapid heating and a high temperature favours the formation of HCN.

The coal gas, as it leaves the retorts, contains the whole of its cyanogen as HCN, which is produced by the action of ammonia on red-hot carbon :—

$$NH_3 + C = HCN + H_2.$$

Now, at a high temperature, steam reacts with HCN, and reconverts it into NH_3, thus :—

$$HCN + H_2O = NH_3 + CO.$$

(See Carpenter and Linder, *Journ. Soc. Chem. Ind.*, 1905, **24**, 63). Consequently, moisture in the coal diminishes the yield of cyanide, and increases that of ammonia.

This is the reason why Mond gas and the gas from coke ovens (where much moisture is present) always contains less HCN than ordinary coal gas.

In general, the gas from coal-gas plants contain more HCN than the gas from coke ovens ; very little HCN occurs in Mond gas.

Consequently the HCN is principally recovered in coal-gas plant, not in coke ovens, as it scarcely pays in the latter case.

There are a great many different methods of treating the gas for cyanide, but we may roughly divide them into two main processes :—

> (1) The old **dry process**, whereby the gas is simultaneously freed both from HCN and H_2S by passing through a dry mass of spent iron oxide (see under **Illuminating Gas** in Martin's "Industrial Chemistry," Vol. I.).
>
> (2) The **Bueb's Wet Process**, whereby the gas is scrubbed with a concentrated ferrous sulphate solution (see Martin's "Industrial Chemistry," Vol. I.).

(1) Dry Process of Prussiate Recovery from Coal Gas.—The spent oxide, as it comes from the gasworks, contains 35-50 per cent. free sulphur, 10-15 per cent. cyanide (in the form of Prussian blue), calculated at present as crystallised ferrocyanide, 1·5 per cent. ammonium sulphocyanide, NH_4CNS, and 1·7 per cent. ammonium sulphate. The substance is valued technically on the cyanogen contents. The mass is usually extracted with water, whereby the NH_4CNS is obtained, together with ammonium sulphate. Sometimes the NH_4CNS is recovered by fractional crystallisation from the dissolved ammonium sulphate, but where this separation is difficult the solution is simply boiled with lime ; the ammonia gas is thus driven out and collected as described under ammonia. The solution is filtered from depositing calcium sulphate, $CaSO_4$, and it then contains the soluble calcium sulphocyanide, $Ca(CNS)_2$. This is treated with ammonium sulphate and converted into ammonium sulphocyanide, thus :—

$$Ca(CNS)_2 + (NH_4)_2SO_4 = CaSO_4 + 2NH_4CNS.$$

The $CaSO_4$ separates out and leaves the sulphocyanides in solution. Sometimes, however, the sulphocyanide is precipitated in the form of the insoluble copper sulphocyanide, $CuCNS$.

The mass remaining after lixiviating with water is usually treated by Kunheim's process of heating with slaked lime, when all the iron cyanide present is transformed into the soluble calcium ferrocyanide, $Ca_2Fe(CN)_6$. The clear filtered solution of calcium ferrocyanide is now treated with KCl, when the difficultly soluble calcium potassium, $K_2CaFe(CN)_6$, separates in the form of small crystals. The equivalent amount of potassium carbonate, K_2CO_3, is added to these, and thus we get formed insoluble $CaCO_3$ and the soluble potassium ferrocyanide, $K_4Fe(CN)_6$. The solution, after filtration, is evaporated, and the potassium ferrocyanide recovered by crystallisation.

The cyanide-free residue, containing much free sulphur, is sold to sulphuric acid manufacturers, who burn the mass (see **Sulphuric Acid**, Martin's "Industrial Chemistry," Vol. II.), and the SO_2 escaping is used for the manufacture of sulphuric acid.

Between the years 1885-1895 the spent oxide of gasworks was the principal source of prussiate, but since 1895 the more efficient wet methods of recovering HCN have been introduced by Bueb, and so the importance of the process has diminished.

In general, when gas is purified by the old process of ammonia scrubbers and iron oxide purifiers, about 33 per cent. of the HCN condenses with the ammonia, and only 50-70 per cent. in the iron oxide purifiers, the rest of the HCN passing away with the coal gas, causing corrosion of the gas holders and meters.

With the more efficient wet processes of cyanogen recovery, whether by **Bueb's process** (below) or by the **sulphocyanide process** (p. 83), practically all the HCN is removed from the coal gas and recovered.

(2) Bueb's Wet Process of Cyanide Recovery from Coal Gas.

—In this process the coal gas, after leaving the retorts and after the deposition of tar, is passed through a saturated solution of ferrous sulphate, $FeSO_4$ (see English Patent, 9,075/98). Over 98 per cent. of the HCN gas is removed from the coal gas (more than three times the amount withdrawn by the "dry process") in the form of a sludge of prussiate, consisting essentially of the insoluble ammonium ferro-cyanide, $(NH_4)_6Fe(Fe(CN_6))_2$, or $2NH_4CN + Fe(CN)_2$, together with some soluble ammonium ferrocyanide, much ammonium sulphate, some ammonium carbonate, and some iron sulphide. The sludge usually contains 15-20 per cent. cyanogen, calculated as being present in the form of crystallised potassium ferrocyanide, $K_4Fe(CN)_6.3H_2O$. The mud is boiled, whereby the soluble cyanide interacts with the FeS present, and is converted into the insoluble ferroammonium cyanide, $2NH_4CN.Fe(CN)_2$. Next the mass is filter-pressed; the liquid, containing much ammonium sulphate, is worked up separately for NH_3. The mass of insoluble cyanide remaining in the press is heated with lime (NH_3 again escaping and being collected), whereby the soluble calcium ferrocyanide, $Ca_2Fe(CN)_6$, is formed. This is then treated with the equivalent amount of sodium carbonate, Na_2CO_3, whereby sodium ferrocyanide, $Na_4Fe(CN)_6$, is formed:—

$$Ca_2Fe(CN)_6 + 2Na_2CO_3 = CaCO_3 + Na_4Fe(CN)_6.$$

The calcium carbonate is filtered off and the clear solution is evaporated for $Na_4Fe(CN)_6$.

By Bueb's process it is possible to directly manufacture sodium ferrocyanide. This is owing to the purity of the calcium ferrocyanide solution. In the case of the spent oxide process, previously described, it is not practical to work the cyanides in the mass directly for sodium ferrocyanide. The spent oxide contains so many impurities that in practice the insoluble $K_2CaFe(CN)_6$ is always first prepared (an analogous sparingly soluble sodium salt, $Na_2CaFe(CN)_6$, does not exist), which is then converted into $K_4Fe(CN)_6$ by treating with K_2CO_3.

Potassium Ferricyanide

$K_3Fe(CN)_6$, or $3KCN.Fe(CN)_3$, contains ferric iron. It is prepared by oxidising a solution of potassium ferrocyanide, $K_4Fe(CN)_6$, ($4KCN.Fe(CN)_2$), by means of chlorine ($2K_4Fe(CN)_6 + Cl_2 = 2K_3Fe(CN)_6 + 2KCl$), or by electrolysis in the presence of some calcium salt, the oxygen coming off at the anode serving as the oxidising agent. The substance crystallises as deep-red monoclinic crystals; 100 g. of water dissolve 36 parts of salt at 10° C., 75.5 at 100° C.

The substance is used as an ordinary agent in dyeing. It also finds some application in the production of blue prints of engineering and other drawings. Paper is coated with a solution of ferric ammonium citrate, or oxalate, and is then exposed to light beneath the transparent drawing. When the light falls on the surface the ferric iron is reduced to the ferrous state. On immersing the print in a solution of potassium ferrocyanide a deposit of Prussian blue is formed on the parts exposed to light, the shaded parts appearing white.

Prussian Blue (Berlin Blue)

is the fine mineral colour containing both ferrous and ferric iron united with the cyanogen radicles. It has been on the market since 1700. Together with ultramarine and the coal-tar blues it is still much used for the manufacture of paint, for paper and cloth printing, etc. For the latter purpose it is often produced on the fibre itself by imprinting with potassium ferrocyanide, followed by steaming (see Vol. I., Martin's "Industrial Chemistry").

Prussian blue is an important colour, since it is fast towards light and acids. Towards bases, however, it is not so fast. Tissues dyed with Prussian blue gradually lose their colour in sunlight, but regain it in the dark. When heated to 170° C. in air it glows and burns away to a brown residue of iron oxide.

Chlorine turns a suspension of Prussian blue in water greenish, but the blue colour is restored by washing with water.

6

Prussian blue is usually regarded as **Ferric ferrocyanide**, $Fe'''_4(Fe''(CN)_6)_3$, or $3Fe(CN)_2 4Fe(CN)_3$, or $Fe_7(CN)_{18}$. The dry substance contains water (which cannot be driven off without decomposition), corresponding approximately to $Fe_7(CN)_{18}10H_2O$.

There is little doubt, however, that the commercial Prussian blue is really a mixture of several different substances, all possessing a blue colour (see Hofmann, Heine, and Höchtlen, **Annalen**, 1904, **337**, 1 ; Hofmann and Resensheck, *ibid.*, 1905, **340**, 267 ; 1905, **342**, 364, for an examination of the whole subject). Among blue bodies which are sometimes present may be mentioned " **soluble Prussian blue**," $2[KFe'''(Fe''(CN)_6)] + 3.5H_2O$. **Hofmann's blue**, $KFe'''(Fe''(CN)_6) + H_2O$. **Stable soluble blue**, $Fe'''K(Fe''(CN)_6) + H_2O$; **Williamson's blue or violet**, $KFe'''(Fe''(CN)_6) + H_2O$. A full account of these different blues is given in Hofmann's papers above referred to.

Manufacture.—Several different varieties of Prussian blue are on the market. The finest commercial Prussian blue goes under the name **Paris blue**, and is made by dissolving 50 kilos of potassium ferrocyanide in 250 kilos of water, and making simultaneously a solution of 43-45 kilos of ferrous sulphate (green vitriol) in 259 kilos of water, best in the presence of scrap iron to avoid formation of ferric salts. The two solutions are now run simultaneously into a vessel containing 250 kilos of water, and the almost white precipitate which forms is allowed to settle and is drained on a cloth filter.

The paste is now heated to boiling, transferred to a wooden vessel, and 25.5 kilos of concentrated HNO_3 (1.23 sp. gr.) mixed with 18 kilos of concentrated sulphuric acid (1.84 sp. gr.) are added. This oxidises the paste to a fine blue colour. After standing twenty-four hours the mixture is suspended in large excess of cold water and settled. The mass is washed with cold water by decantation until the bulk is free from sulphuric acid. It is then collected on linen filters, pressed to cakes, and dried in air at 39°-40° C. Yield, 39-40 kilos.

Another method is to acidify the paste with hydrochloric acid and pass chlorine gas through it, until the solution gives a deep blue colour with potassium ferrocyanide.

Still a third method is to oxidise ferrous sulphate with nitric acid and run into the solution potassium ferrocyanide solution. The deep blue precipitate is collected and washed until free from iron.

The best qualities of Prussian blue (known as Paris blue) are made as above described. But inferior qualities (known as **Mineral blue**), are sold in which the Paris blue is mixed with starch, gypsum, burnt and finely-ground kaolin, heavy spar, etc.

The mixture is made by adding the white finely-ground material to the Paris blue paste, and passing through a colour mill.

Sulphocyanides or Thiocyanates—Recovery from Coal Gas.— Sulphocyanides are now obtained solely from coal gas. At one time synthetic sulphocyanides were made, but apparently the manufacture has been abandoned.

Gas liquor, when quite fresh, contains ammonium sulphide, $(NH_4)_2S$, and ammonium cyanide, NH_4CN. When stored, the atmospheric oxygen sets free sulphur, which dissolves to form polysulphides, and these then react with the cyanide present to form sulphocyanide, thus :—

$$(NH_4)_2S_2 + NH_4CN = NH_4SCN + (NH_4)_2S.$$

By means of this reaction, Wood, Smith, Gidden, Salamon, and Albright (English Patent, 13,658, 1901) introduced a successful process of recovering practically all the cyanogen (98 per cent.) of coal gas, which is now worked by the British Cyanides Co. Ltd., and by several gas works. It yields practically the whole of the sulphocyanides now made.

The coal gas coming from the retorts is first led through a tar extractor, and then through a special scrubber, after which it passes on to the ammonia scrubber of the works. The water in this special scrubber is rendered ammoniacal by the passage of the gas, and sulphur in small lumps is then added.

As the sulphur is rotated by the scrubber, polysulphides of ammonia are formed, which are in turn decomposed by the hydrocyanic acid, HCN, of the coal gas, with the production of sulphocyanide of ammonia.

The ammonium sulphocyanide, NH_4CNS, is next distilled with lime, and the ammonia is expelled and recovered, thus :—

$$Ca(OH)_2 + 2NH_4CNS = Ca(CNS)_2 + 2NH_3.$$

The calcium sulphocyanide, $Ca(CNS)_2$, is next converted into sodium or potassium sulphocyanide by treating with sodium or potassium carbonate :—

$$Ca(CNS)_2 + Na_2CO_3 = CaCO_3 + 2NaCNS.$$

The sodium or potassium sulphocyanide can then be converted either into cyanide or into prussiate (ferrocyanide), as described on p. 75.

In 1909 P. E. Williams, engineer at Poplar, designed a special purifier box, which merely contains moistened spent oxide—about 50 per cent. S. The box arrests sulphur, in some way which is not yet quite clear, and it is claimed that the CS_2 in the gas is also reduced. The saving effected over the former method, using a mechanically driven washer fed with specially prepared sulphur, must be considerable.

It will thus be seen that the recovery of cyanogen from coal gas is either effected as ferrocyanide, as described on p. 80, or as sulphocyanide. The *advantage of the ferrocyanide process* is that it gives a product which itself has a fairly wide market, and can be easily transformed into either Prussian blue or alkali cyanides. Its disadvantage is that it is almost impossible—with the intentional admission of a small proportion of air to the coal gas before purification in order to effect some continuous revivification of the oxide, *quite* impossible—to avoid the formation of sulphocyanides during the process of purification. The recovery of this where a ferrocyanide process is worked is usually too costly, and it is generally lost. The great *disadvantages of the sulphocyanide processes* are the facts that it has not been found practicable to manufacture alkali cyanides direct from sulphocyanide and that the sulphocyanide liquor is very corrosive to wet iron.

SYNTHETIC SULPHOCYANIDES FROM CARBON DISULPHIDE AND AMMONIA

At one time considerable amounts of synthetic sulphocyanides were made and worked for cyanides or sulphocyanides, as described above. At the present time the process appears to have been abandoned, but may later revive, and so we will describe the process.

Gelis (English Patent, 1,816, 1860) manufactured sulphocyanide by agitating a concentrated solution of ammonia and ammonium sulphide with carbon disulphide, CS_2, when ammonium sulphocarbonate is formed, thus :—

$$(NH_4)_2S + CS_2 = (NH_4)_2CS_3.$$

The solution is next heated to 90°-100° C. with potassium sulphide, when potassium sulphocyanide is formed with the evolution of much sulphuretted hydrogen :—

$$2(NH_4)_2CS_3 + K_2S = 2KCNS + 2NH_4HS + 3H_2S.$$

The disposal of the sulphuretted hydrogen evolved caused much expense, and the process was not successful commercially. Günzburg and Tcherniac (see English Patents, 1,148, 1878, 1,359, 1879, 1,261, 1881) improved the process by simply heating a 20 per cent. NH_3 solution to 100° in an autoclave provided with stirrers, until the pressure of 15 atmospheres was reached, when ammonium sulphocyanide is quantitatively produced, thus :—

$$4NH_3 + CS_2 = NH_4SCN + (NH_4)_2S.$$

The process, also, was not commercially successful. In 1894, however, Brock, Raschen, and others (see Crowther and Rossiter, English Patent, 17,846, 1893; Bock, Hetherington, Hurter, and Raschen, English Patent, 21,451, 1893) modified the process by adding lime to the charge, thus diminishing the pressure in the autoclaves and reducing the amount of ammonia required, the action taking place according to the equation :—

$$2NH_3 + 2CS_2 + 2Ca(OH)_2 = Ca(SCN)_2 + Ca(SH)_2 + 4H_2O.$$

Excess of ammonia must be present to prevent formation of sulphocarbonate. At the end of the process this is distilled off, CO_2 gas is driven through the liquid to expel the sulphuretted hydrogen (as in the Claus-Chance treatment of alkali waste), and any precipitated calcium carbonate is filtered off. The soluble calcium sulphocyanide is treated with sodium carbonate and converted into sodium sulphocyanide, thus :—

$$Ca(SCN)_2 + Na_2CO_3 = CaCO_3 + 2NaSCN.$$

The United Alkali Co. worked this process for some years, converting the synthetic sodium sulphocyanide into cyanide by Raschen's process of oxidation by nitric acid, as described on p. 75.

CHAPTER VIII

MANUFACTURE OF
NITROUS OXIDE
(Laughing Gas, Nitrogen Monoxide), N_2O

THE usual process is to heat ammonium nitrate, NH_4NO_3, in retorts, when it decomposes, thus :—

$$NH_4NO_3 = N_2O + 2H_2O.$$

The substance begins to decompose at 170° C., and the heat must be carefully regulated (best by gas firing), otherwise explosions can occur.

It is important to use pure ammonium nitrate. If the temperature is too high, N, NH_3 and the very poisonous NO are produced. The gas must be purified by passing through solutions of ferrous sulphate, $FeSO_4$, caustic potash KOH, and milk of lime. NO is caught by the $FeSO_4$; and is held back by the KOH and lime, which also retains any CO_2.

·1 kilo NH_4NO_3 gives 182 l. N_2O.

See Baskerville and Stephenson (*Journ. Ind. Eng. Chem.*, 1911, **3**, 579) for a full account of its preparation and the requisite purity for use as an anæsthetic.

Lidoff (*Journ. Russ. Phys. Chem. Soc.*, 1903, **35**, 59) mixes the ammonium nitrate with sand, and washes the gas with ferrous sulphate solution, drying it with an emulsion of ferrous sulphate in concentrated sulphuric acid.

Smith and Elmore (D. R. P. 71,279 of 1892) heat dry KNO_3 with dry $(NH_4)_2 SO_4$. The evolution of gas begins at 230° C. and ends at 300° C. Thilo (*Chem. Zeit.*, 1894, **18**, 532) uses $NaNO_3$ and heats to 240° C. Campari (*Chem. Cent.*, 1888, 1569) heats 5 parts $SnCl_2$, 10 parts HCl (sp. gr. 1.21), and 0.9 parts HNO_3 (sp. gr. 1 38), when gas is evolved. Pictet (French Patent, 415,594 of 1910,) and Södermann (French Patent 411,785 of 1910) obtain it from an electrically produced nitrogen-oxygen flame by rapid cooling.

Properties.—Colourless gas with pleasant odour and sweet taste. Density, 1.5301 (Air = 1). 1 l. weighs 1.9774 g. at 0°C. and 760 mm. Coefficiency of expansion, 0.0037067.

The liquefied gas has density 1.2257 ($H_2O = 1$), and refractive index 1.193 at 16°C. Critical temperature, 35.4° C.; critical pressure, 75 atmospheres. Liquid boils at –88° C, and thereby partially solidifies (at –115° C.). Mixed with CS_2 evaporated in vacuo, a temperature of –140° C. is attained.

Burning oxidisable bodies (such as P, S, etc.) continue to burn in the gas as in pure oxygen. With H gas it forms an explosive mixture. Heat of formation, 21,700 calories.

Solubility in water :—

1 volume water dissolves	5° C.	10° C.	15° C.	20° C.	25° C.
	1.048	0.8778	0.7377	0.6294	0.5443 vols.

When breathed, nitrous oxide is a valuable anæsthetic for short operations ; 22-26 l. of gas are needed to produce insensibility. Prolonged breathing causes death. It is advisable to mix 0.1 per cent. of atmospheric air with the gas ; the limit is 0.25 per cent. air. Mixed with oxygen the breathing of the gas produces intoxication. Also used for making **sodium azide** (see p. 106).

Analysis.—Best by burning with H according to Bünsen's method. See W. Hempel (B. **15**, 903 ; *Journ. Soc. Chem. Ind.*, 1882, 200). A. Wagner (*Journ. Soc. Chem. Ind.*, 1882, 332) gives a method of estimating the NO in N_2O. See also Lunge, *Journ. Soc. Chem. Ind.*, 1881, 428.

Transport.—In liquified form in iron, steel, or copper cylinders. See R. Hasenclever, *Chem. Ind.*, 1893, 373.

CHAPTER IX

—

The Modern Explosives Industry

THE MODERN EXPLOSIVES INDUSTRY

LITERATURE

"Treatise on Service Explosives."

BRUNSWIG.—"Explosivstoffe."

BERTHELOT.—"Explosives and their Power." Translated by Hake and Macnab.

Sir A. NOBEL.—"Artillery and Explosives."

COOPER-KEY.—"A Primer of Explosives."

CUNDILL AND THOMSON.—"Dictionary of Explosives." 1895.

EISSLER.—"Handbook of Modern Explosives." 1897.

WALKE.—"Lectures on Explosives." U.S.A., 1897.

GUTTMANN.—"Manufacture of Explosives." 1895.

WESSER.—"Explosive Materials."

SANFORD.—"Nitro-Explosives." 2nd Edit. 1906.

BICHEL.—"New Methods of Testing Explosives." 1905.

GUTTMANN.—"Twenty Years' Progress in Explosives." 1909.

Dr RICHARD ESCALES.—"Explosivstoffe": "Das Schwarzpulver," 1904; "Die Schiess-baumwolle," 1905; "Nitroglyzerin und Dynamit," 1908; "Ammonsalpetersprengstoffe, 1909; "Chloratsprengstoffe," 1910.

"Improvements in Production and Application of Gun-cotton and Nitro-glycerine."—Sir FREDERIC L. NATHAN's paper at the Royal Institution. See *Nature* for 1909, pp. 147 and 178; also *Journ. Soc. Chem. Ind.*

"Gun-cotton Displacement Process." *Zeitschrift für das Gesamte Schiess- und Sprengstoffwesen*, 1906, I, p. 1.

"Acetone Recovery at the Royal Gunpowder Factory." *Loc. cit.*, 1910. 5, 446.

COOPER.—"The Rise and Progress of the British Explosives Industry." 1909.

ALLEN's "Organic Analysis." New Edition. 1910.

Special points may be found dealt with in the *Journals* as indicated in the Article.

MODERN EXPLOSIVES

Gunpowders

THE term "modern," as applied to explosives, is generally understood to connote an explosive other than black gunpowder or the later brown powder. Gunpowder, already known in the middle of the thirteenth century, is still used, however, for sporting purposes and for mining, though it has been discarded by military authorities in favour of the more powerful agents now at our disposal. Its constituents, saltpetre, sulphur, and charcoal, are still mixed in the same way and in the same proportions as formerly (75 per cent. KNO_3, 10 per cent. S, 15 per cent. C).

In this country, **Bobbinite**, essentially a black powder with diminished sulphur content, headed the list of "permitted" explosives arranged in order of quantities used during 1909. Although a Departmental Committee, appointed in 1906 by the Home Office to investigate allegations as to danger arising from the use of Bobbinite in fiery mines, reported that its use should not for the present be restricted, recent Explosives in Coal Mines' Orders state that it is permitted only for the purpose of bringing down coal in certain mines for a period of five years from 1st January 1914. It is made in two varieties: (*a*) KNO_3, 62-65 parts; wood charcoal, 17-19.5 parts; sulphur, 1.5-2.5 parts; together with 13 to 17 parts of a mixture of ammonium sulphate and copper sulphate; or (*b*) KNO_3, 63-65 parts; wood charcoal, 18.5-20.5 parts; sulphur, 1.5-2.5 parts; rice or maize starch, 7-9 parts, and 2.5-3.5 parts of paraffin wax.

H.M. Inspector's report shows that during 1909, out of 30,091,887 lbs. of explosives used in mines and quarries, 17,595,475 lbs. were gunpowder. This constituted 58.5 per cent. of the whole. Gunpowder is not a "permitted" explosive, and in Germany, also, its use in coal mines is forbidden, the prohibition being extended to all powders of the same type.

The great developments of the potash industry in Germany have involved a revival of mixtures resembling black powder, and these have a large sale. Both in that country and in America nitrate of soda, on account of its cheapness, is substituted for potassium nitrate, and charcoal is frequently replaced by brown coal or pitch. **Sprengsalpeter**, for instance, consists of 75 per cent. $NaNO_3$, 15 per cent. brown coal, and 10 per cent. sulphur. Owing to the hygroscopic nature of the sodium nitrate, the mixture of this substance with the carbonaceous material and sulphur is heated to the melting point of sulphur and then pressed, the resulting product being only slightly affected by moisture. Explosives of this type come into commerce either in the form of grains packed in cartridges or as compressed cylinders with an axial canal, and are fired in the same way as black powder. They burn slowly when ignited and fissure the rock rather than shatter it. **Petroklastite** is an explosive of this class and contains: $NaNO_3$, 69 per cent.; KNO_3, 5 per cent.; S, 10 per cent.; pitch, 15 per cent.; $K_2Cr_2O_7$, 1 per cent. The introduction of ammonium nitrate secures a lower temperature during explosion. **Wetter-Dynammon**, made by the Austrian Government for use in fiery mines, contains: NH_4NO_3, 93.83 per cent.; KNO_3, 1.98 per cent.; charcoal, 3.77 per cent.; moisture, 0.42 per cent.; and **Amide powder** has the composition: KNO_3, 40 parts; NH_4NO_3, 38 parts; charcoal, 22 parts.

Gunpowder, again, is still used in priming compositions in order to initiate the decomposition of high explosives, and in certain types of safety fuse it continues to hold its own. The latter may be compared to a piece of cord having a fine train of gunpowder down the centre. Stress of commercial competition has resulted in the production of a greatly improved fuze which burns uniformly through a definite length in a given time. Attempts to substitute a core of smokeless powder for gunpowder have not as yet yielded a satisfactory result.

Chili saltpetre (sodium nitrate) contains perchlorates, as also does "conversion" saltpetre or potassium nitrate prepared from it. To this circumstance, when discovered, was attributed the origin of various disastrous explosions which had previously occurred, but later work has shown that, in ordinary cases, the danger arising from the presence of a small percentage of the more sensitive perchlorate has been over-estimated (Dupré, *Journ. Soc. Chem. Ind.*, 1902, 825; Lunge and Bergmann's paper read at Fifth International Congress of Applied Chemistry, Berlin, 1903). Bergmann recommends a maximum of 0.4 per cent. of perchlorate for blasting powders and 0.2 per cent. for gunpowders.

Nitro-glycerine, the trinitric ester of glycerol, $C_3H_5(ONO_2)_3$, is an important constituent of a great many explosives used as propellants or for blasting. Should the price of glycerol continue at its present high figure it is not improbable that explosives made from cheaper raw materials will displace it to some extent.

The process of manufacture now about to be described is the outcome of a long series of improvements which have been made at intervals upon the methods originally adopted by Alfred Nobel, the pioneer in the commercial production of this substance. The nitrator-separator in its present form was first used at Waltham Abbey in the Royal Gunpowder Factory, and is the subject of the British Patent 15,983 of 1901, taken out by Colonel Sir Frederic L. Nathan, R.A., Mr J. M. Thomson, F.I.C., and Mr Wm. Rintoul, F.I.C.

The apparatus is shown diagrammatically in the accompanying Fig. 28 in both plan and elevation. The nitrating vessel a has the form, generally used, of a lead cylinder with sloping bottom and conical lead roof. The mixed acids are not introduced from above, as has been customary, but through a pipe connected with the lowest part of the bottom. This pipe has three branches, b, c, and d; b leads to the denitrating plant, c to the drowning tank, while d ascends and divides into two branches of which one is connected with the storage tank for mixed acid and the other to the waste acid tank. The acid pipe bend is one foot lower than the lowest part of the bottom so that, during stirring, no nitro-glycerine may enter the ascending pipe. On the sloping bottom of the vessel lies a coil g provided with several holes for the introduction of compressed air for stirring, and in the cylindrical portion of the vessel are cooling coils $h\,h$ of lead pipe through which water, refrigerated if necessary till its temperature is below 10° C., flows. The conical roof is surmounted by a lead cylinder e with two windows f and side outlet, which is connected with the lead pipe k down which the nitro-glycerine flows to the pre-wash tank. From the pipe k ascends another pipe m, through which fumes are aspirated. s is a thermometer, the bulb of which dips into the mixture inside a.

To carry out a nitration, the mixed acids, consisting of 41 per cent. HNO_3, 57.5 per cent. H_2SO_4, and 1.5 per cent. H_2O in the ratio of 6.09 of mixed acid to 1 of glycerine, are allowed to enter through d, air-stirring is begun, and the glycerine is introduced by means of an injector, the mixture being cooled by the coils $h\,h$. The glycerine, owing to its viscosity at ordinary temperatures, must have been previously warmed to 30°-35° C., and its rate of admission is regulated so as to prevent the temperature from rising above 22° C. When nitration is finished, and the temperature has fallen a little, the air in *g* and the water in *h* are shut off, and the mixture then begins to separate.

The temperature should not have been allowed to fall below 15° C. Waste acid from a previous operation is now allowed to enter the apparatus slowly from beneath, through *d*, and forms a layer of gradually increasing depth below the reaction mixture which thus rises slowly in the vessel. The nitro-glycerine reaches the cylinder *e*, and runs off through *k* to the pre-wash tank, containing water, and is subjected to an alkaline washing. The displacing acid enters at such a rate as to cause clear nitro-glycerine, free from suspended acid, to flow through the outlet as separation proceeds. Finally a sharp line of demarcation between acid and nitro-glycerine is observed through *f*, and the inflow of acid through *d* is stopped. The contents are allowed to stand for a time in order to obtain the best possible yield of nitro-glycerine.

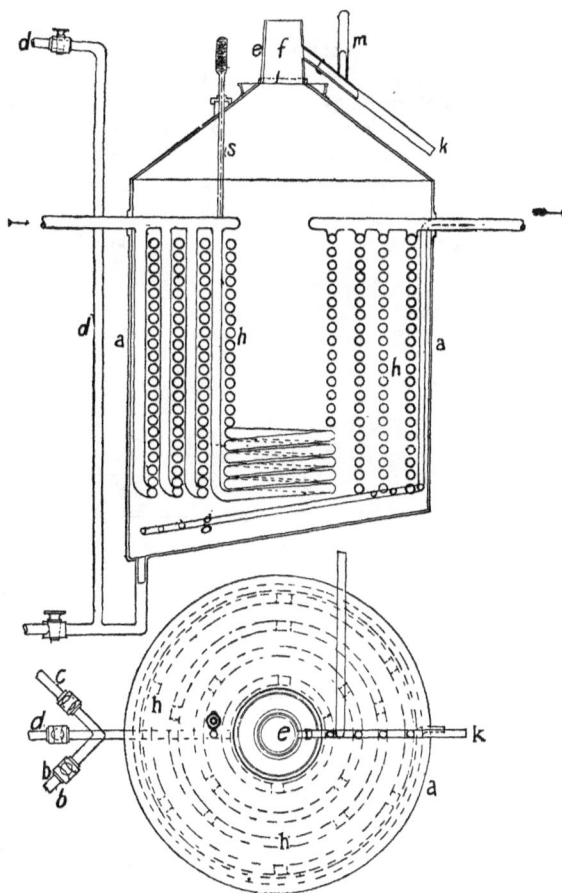

FIG. 28. —Nathan, Thomson, and Rintoul's Nitrator-Separator.

Under the old régime the waste acid was run off to a special building, the after-separating house, in which nitro-glycerine still remaining in the acid separated out slowly during the course of several days. This was skimmed off from time to time, and put into a vessel filled with water in which it was washed. A workman was required to be in attendance day and night. The modern practice is to destroy the nitro-glycerine in the waste acids immediately after nitration by the following method, due to Sir F. L. Nathan and Messrs Thomson and Rintoul :—

Sufficient waste acid is run off from the nitrator-separator to avoid splashing over during stirring, which is again started by turning on the compressed air. Water amounting to 2 per cent. of the waste acid is slowly added, and mixes immediately with the acid through the stirring. The temperature rises slowly during the addition of the water, about 3° C. for each per cent. of water added, and air stirring is continued after the addition of water till the temperature begins to sink. The air is then shut off, and waste acid again run in from beneath till the surface reaches the upper outlet. On standing for a time nitro-glycerine adhering to the sides of the vessel and the cooling coils may ascend, and can be run off through *k*. The waste acid, now completely free from nitro-glycerine, is allowed to run off through *b*, as much as will be necessary for displacement of nitro-glycerine in a later operation being elevated to the storage tank communicating with *d*, while the remainder is run into a tank where it stays without further supervision until it is denitrated.

Owing to the introduction of this process the nitro-glycerine never passes through an earthenware cock at any stage of its manufacture and a source of danger is thereby removed. A further lessening of risk is achieved by the removal of the nitro-glycerine immediately after its separation from the acid, by the cooling during separation and after separation, by decrease in amount of manipulation of nitro-glycerine and waste acid, and by the disappearance of the necessity for an after-separation building. The saving in differences of level necessary to transport the nitro-glycerine from building to building is very considerable, and the area necessary for the factory is now less than formerly. A plant consisting of two nitrating vessels and one pre-wash tank nitrates five charges each of 1,450 lbs. of glycerine daily.

After removal of acid from the nitro-glycerine in the pre-wash tank, accomplished by air-stirring with water and finally with an alkaline solution, the nitro-glycerine is allowed to run through a rubber pipe attached to the lowest point of the sloping bottom of the pre-wash tank into the gutter, which has a fall of 1 in 60 and leads to the final washing house. During the preliminary washing the rubber pipe is secured to a nozzle fixed to the outside of the tank at a point above the level of the liquid.

In the final washing house the nitro-glycerine is subjected to prolonged treatment with alkaline solution, using air-stirring as before. The alkali remaining after this treatment is removed by washing with pure warm water, and the nitro-glycerine is afterwards filtered by passing it through a mat of sponges in order to free it from small quantities of flocculent impurities and suspended mineral matter. A diagram of the plant is shown in the figure (Fig. **29.**)

By the use of the acid mixture above mentioned, obtained from 2.8 parts of nitric acid and 3.4 parts of fuming sulphuric acid (20 per cent. SO_3), Sir F. L. Nathan and Mr Wm. Rintoul have,

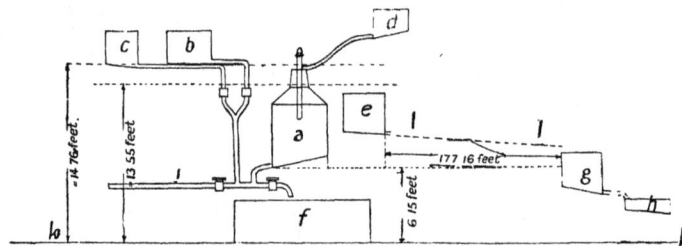

FIG. **29.** —Arrangement of a Modern Nitro-glycerine Mill.

A. The Nitrator-Separator	G. Final Washing Tank
B. The Mixed Acid Tank	H. Filter
C. Waste Acid, for Displacement Tank	I. Pipe for running off waste Acid
D. Glycerine Tank	K, K. Ground Level
E. Pre-wash Tank	L, L. Gutter, with fall of 1 in 120
F. Drowning Tank	

using the above process, increased the yield of nitro-glycerine to 230 parts per 100 parts of glycerine (British Patent, 6,581, 1906). By former processes in which 100 parts of glycerine were added to a mixture of 300 parts of 94 per cent. HNO_3 and 500 parts of 96 per cent. H_2SO_4, the yields varied from 210-215 per cent. Using this acid mixture in the Nathan, Thomson, and Rintoul process the yield was increased to 220 per cent., the theoretical yield being 246.7 per cent.

The use of waste acid from a previous nitration which is brought to its original composition by the addition of fresh acid gives, according to data from the Zentral-stelle, Neu Babelsberg, yields of 228 per cent. (British Patent, 2,776, 1905). Details regarding the influence of mixed acids which have been thus revivified upon the time necessary for complete purification of the nitro-glycerine are not available.

The process of separation, after nitration, is frequently protracted owing to the action of gelatinous silicic acid in preventing the nitro-glycerine from passing from the emulsified form to that of a separate layer of pure liquid. In order to obtain a quick and well-defined separation the Dynamit A. G. (D.R. Patent, 171,106, 1904) at Hamburg, add, during nitration, paraffin or fatty acids of high molecular weight in proportions of 0.5-2 per cent. by weight of the glycerine, while Reese (U.S.A. Patent, 804,817, 1905) uses .002 per cent. of sodium fluoride, whereby the silicic acid is converted into silicon tetrafluoride.

After-separation has been shortened, on the laboratory scale, by Escales and Novak (British Patent, 18,597, 1907) by the use of the electric current, but is entirely obviated in the Nathan,

Thomson, and Rintoul process. Evers (D.R. Patent, 182,216) has described a new process for denitrating waste acid, in which a mixture of air and steam at a high temperature enters the denitrating tower at one or more points.

Properties of Nitro-glycerine.—Kast has found that nitro-glycerine has two solidification or melting points (*Zeitschr. Schiess-Sprengstoffwesen*, i. 225). The lower melting phase is a labile product and changes completely into the stable higher melting modification on stirring with a crystal of the latter, which melts at about 13° C. The thawing of frozen nitro-glycerine has been a fruitful source of accident, and attention has been directed towards obtaining a product which will remain liquid under the conditions to which it is subjected during cold weather. The well-known effect of adding soluble ingredients is to produce a **depression of freezing point** proportional to the amount dissolved, and nitro-benzene and ortho-nitrotoluene were used for this purpose. Other nitric esters of glycerine, however, which do not appreciably diminish the explosive power of the product to which they are added, have recently found extended use for this purpose.

Dinitro-glycerine (glycerine dinitrate), first prepared by Mikolajzak in 1903 for use in depressing the freezing point of trinitro-glycerine (British Patent, 27,706, 1904), may be prepared alone, and then added to the latter, or, more conveniently, the two esters may be generated simultaneously under suitable conditions in the proportions required.

One method of working (Heise, "Der Bergbau," 1907, Heft 35) is to nitrate the glycerine in nitric acid, allowing the temperature to rise to a certain extent; the solution of the ester in nitric acid is then allowed to stand for about two hours, and separation of the ester is then effected by diluting with water and neutralising with carbonate of lime. After separation, the dinitro-glycerine still contains water and acid, the latter being removed by adding soda and the water by a special process. A reddish yellow oil of sp. gr. 1.5 is thus obtained. The calcium nitrate solution obtained as a bye-product is treated with ammonium sulphate; the ammonium nitrate formed is used in explosives manufacture, and the sulphate of lime as a manure. Extraction of the diluted acid mixture, both before and after neutralisation, by means of ether, has been tried on the small scale, and Escales and Novak have proposed glycerine disulphuric ester as the starting point for the preparation. The method of preparing a mixture of di- and trinitro-glycerine is described by Mikolajzak (*loc. cit.*).

The technical product is a mixture of two isomeric modifications. Its purification is difficult.

According to Will (*Berichte*, 1908, 1107; *Zeitschr. Schiess-Sprengstoffwesen*, 1908, 324) the dinitrate, in the water-free condition, is not solid at ordinary winter temperatures, but offers little advantage over nitro-glycerine from the point of view of sensitiveness to shock. It is hygroscopic, and when moist is less sensitive, but at the same time loses its faculty for forming no solid on subjection to winter temperatures.

Nitrate of Polymerised Glycerine.—By appropriate heat treatment 2 molecules of glycerine condense to form 1 molecule of diglycerine with splitting off of water. By treating the diglycerine so formed with mixed nitrating acids, tetra-nitro-diglycerine, $\begin{matrix} C_3H_5(ONO_2)_2 \\ C_3H_5(ONO_2)_2 \end{matrix}\Big\rangle O$, is obtained, which behaves like trinitro-glycerine as regards solubility in organic solvents, insolubility in water, and ease of detonation with fulminate of mercury. The addition of 20-25 per cent. of the product to trinitro-glycerine will prevent it from freezing at ordinary winter temperatures.

Will and Stöhrer (*Zeitschr. Schiess-Sprengstoffwesen*, i. 231) polymerised glycerine by heating it for seven or eight hours to 290°-295° C., and obtained a product containing 60 per cent. diglycerine and 4-6 per cent. of tri- or polyglycerines. These were separated by fractional distillation under reduced pressure. By repeated treatment pure diglycerine can be obtained as a clear, viscous, sweet liquid, soluble in water, having a sp. gr. of 1.33, and distilling undecomposed at 240°-250° C. under 8 mm. pressure.

Escales and Novak treat glycerine with hydrochloric acid and nitrate the resulting mixture of monochlorhydrin, dichlorhydrin, diglycerine, chlorhydrin of diglycerine, triglycerine and chlorhydrin of triglycerine (*Deutsche patentanmeldung*, 12th May 1906).

Claessen (British Patent, 9,572, 1908) produces diglycerine and polyglycerines in a much shorter time than usual by heating the glycerine with a small quantity of alkali (0.5 per cent.) to 275°-280° C. The glycerine, diglycerine, and polyglycerine can be separated by fractional distillation.

Dinitromonochlorhydrin.—By passing hydrochloric acid gas into glycerine heated to 70°-100° C., a mixture of α- and β-monochlorhydrins is obtained

$[CH_2Cl.CH(OH).CH_2(OH)]$ and $[CH_2(OH).CHCl.CH_2(OH)]$. These are separated from the reaction mixture by fractional distillation under reduced pressure, and are afterwards nitrated alone or in admixture with glycerine. In the latter case the reaction is claimed to be less dangerous, and the separation quicker (Welter, British Patent, 6,361, 1905). Dinitromonochlorhydrin is easily stabilised in the usual way, and stands the Abel heat test for a much longer time than ordinary nitro-glycerine. It is a pale yellow liquid with a faint aromatic smell, soluble in ether, alcohol, acetone, chloroform, etc., and insoluble in water and acids. Its sp. gr. is 1.5408 at 15° C., and it boils at atmospheric pressure at 190°-193° C. If boiled under a pressure of 15 mm., no partial decomposition takes place, and it distils over at 120°-123° C. as a colourless oil.

It is surprisingly insensitive to shock and friction, is not hygroscopic, does not freeze at − 25° to − 30° C., and dissolves easily in nitro-glycerine. The addition of 20 per cent. to nitro-glycerine renders it practically unfreezable. It is said to be an excellent gelatinising medium for nitro-cellulose. One disadvantage it possesses, viz., the formation of HCl on explosion, but this can be rectified in large measure by adding an alkali nitrate to the explosive.

Di-nitro-acetin and **Di-nitro-formin** are produced by the nitration of mono-acetin, $C_3H_5(OH)_2.O.CO.CH_3$, and monoformin, $C_3H_5(OH)_2.O.CO.H$. Their properties resemble those of dinitromonochlorhydrin, and they have been used for making unfreezable nitro-glycerine explosives (see *Zeitschr. Sch.-Sprengstoffwesen*, 21, 1907).

Gun-cotton

The process now coming into general use for gun-cotton manufacture is the result of important improvements devised by Messrs J. M. and W. T. Thomson of the Royal Gunpowder Factory, Waltham Abbey (British Patent, 8,278, 1903; D.R. Patent, 172,499, 1904). The following description is taken largely from the paper by Colonel Sir F. L. Nathan in the *Journal of the Society of Chemical Industry*, 27th February 1909. To the Council of the Society and to Sir Frederic Nathan my best thanks are due for their kind permission to use the accompanying illustrations.

The raw materials are cotton waste and sulphuric and nitric acids. Ordinary cotton waste varies considerably in character, and this is not without effect on the character of the resulting gun-cotton. Before it is received at the works it has been subjected to treatment with alkaline, bleaching and acid liquors in succession, after which it has been washed and dried; or the oil, etc., has been extracted with benzol. The physical condition of the cotton, and, to some extent, its chemical properties, are affected by this process, and an undue proportion of "fly" or finely divided and felted material will hinder the penetration of mixed acids during nitration and diffusion of waste acid from the sphere of reaction, and may indicate the presence of products of hydrolysis or of oxidation of cellulose. Oxy-cellulose exerts an influence on the nitrogen content of the gun-cotton and on its solubility in ether alcohol if it be present in appreciable quantity. Hydrocellulose, produced by hydrolytic action, yields a nitro-cellulose agreeing in properties with that obtained from untreated cellulose so far as nitrogen content and solubility in ether alcohol are concerned. The viscosities of nitro-hydrocellulose solutions in acetone are less than those of similar nitro-cellulose solutions, however, and this fact taken in conjunction with the power of reducing Fehling's solution and of fixing methylene blue, possessed by hydrocellulose, indicates a structural change which may possibly affect adversely the stability of gun-cotton containing its esters.

Blotting-paper made from cotton, paper shreds, tissue paper, cellulose as used for paper making, and other raw materials have been suggested for the manufacture, but all have been discarded, and even celluloid and artificial silk makers use cotton, though paper is still largely used in the manufacture of celluloid.

The prepared cotton waste on arrival at the factory contains a notable percentage of hygroscopic moisture, also wood, pieces of iron, metal, string, rubber, etc. The mechanical impurities are removed as far as possible by hand-picking, and the cotton is then passed through a teasing machine (Fig. 30), which separates the cotton fibres and opens out knots and lumps. It is then again picked over as it leaves the machine.

At Waltham Abbey the cotton, as it leaves the teasing machine, is delivered on to an endless band which carries it to the drying machine. The moisture is expelled from the cotton by a blast of hot air supplied by a fan through a steam heater. The cotton waste passes very slowly through the machine, the operation lasting about three-quarters of an hour, and issues from it containing about $\frac{1}{2}$ per cent. of moisture. It is immediately weighed out into charges and placed in sheet-iron boxes or other suitable receptacles, with lids, to cool, for which a period of about eight or nine hours is sufficient. During the cooling the cotton waste reabsorbs about half a per cent. of moisture.

The nitric and sulphuric acids employed are commercial products of a fairly high standard of purity. The waste acids are revivified to a larger extent than formerly, by the use of Nordhausen sulphuric acid and concentrated nitric acid, and are used again for nitration.

The nitrating apparatus consists of a number of units of four pans worked together, Figs. 31 and 32. The pans are of earthenware and circular, 3 ft. 6 in. in diameter and 10 in. deep at the side of the pan; the pan has a fall of 2 in. to the outlet, which is $\frac{3}{4}$ in. in diameter, and is supported on earthenware pedestals about 1 ft. 10 in. above the floor level. The four members of each unit are connected together by lead pipes, and these again are connected to the nitrating acid supply pipe, to the strong and weak waste acid pipes, and to a waste water pipe, through a gauge-box where the rate of flow is determined whilst the waste acids are being run off.

The process proceeds as follows : A small perforated plate is placed over the outlet of each pan, and four perforated segment plates making a complete disc of diameter about 1 in. less than that

FIG. 30. —The Teasing Machine.

of the pan are placed on the bottom. Aluminium fume hoods, which are connected to an exhaust fan, having been placed on the four pans, nitrating acid, which is cooled in summer and warmed in winter, is allowed to rise to the proper level in the pans. There are then 600 lbs. of mixed acids above the bottom plates and 50 lbs. below. A charge of 20 lbs. of cotton waste is then immersed in the acid, handful by handful, aluminium dipping forks being used for the purpose. When all the cotton waste has been pushed under the surface of the acid, perforated plates, in segments, are placed on the top of it, care being taken that all cotton waste is below the surface of the acid, and a film of water, at a temperature of $5°$-$8°$ C., is run very gradually on to the surface of the plates through a distributor. The film of water prevents the escape of acid fumes and the fume hoods are then removed. The time required for dipping a charge is a quarter of an hour. Nitration is allowed to proceed for two and a half hours. The cock leading to the gauge-box is then opened and waste acid is run off at the rate of about 17 lbs. a minute. Water, cooled if necessary, is run on to the top of the perforated plates, through the distributor, at an equivalent rate. The major portion, amounting to about 80 per cent. of the total waste acid, is returned to the acid store tanks to be revivified with Nordhausen sulphuric and new nitric acids. The remaining 20 per cent. of the waste acid is sent to the acid concentration factory for denitration and concentration. The quantity of acid thus dealt with amounts to about 4 lbs. for every pound of gun-cotton.

	H_2SO_4.	HNO_3.	HNO_2.	H_2O.
Composition of waste acid to be revivified (=80 per cent. of total waste acid) - - - -	72.70	17.30	0.65	9.35
Composition of waste acid to be denitrated and concentrated (=20 per cent. of total waste acid) -	61.0	17.35	0.55	21.10

WALTHAM ABBEY GUNCOTTON DISPLACEMENT PLANT

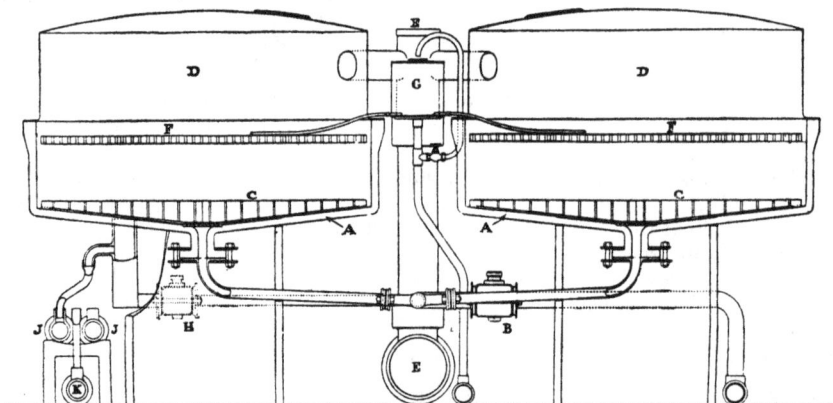

FIG. 31.

A small proportion of the water which follows the recoverable waste acid is slightly acid to the extent of 0.1 lb. for every pound of gun-cotton made, and this represents the total loss of acid (=0.4 per cent.). Mr G. W. MacDonald (*Journ. Soc. Chem. Ina.*, 20th February 1911) finds that the sulphuric acid is displaced at a relatively more rapid rate than the nitric acid.

Stabilisation or thorough purification of the gun-cotton next follows. Adsorbed acids should not be neutralised at an early stage of the purification process, as their hydrolytic action upon various impurities is an important factor at this juncture.

FIG. 32. —Gun-cotton Nitrating House, Waltham Abbey.

Amongst the impurities to be removed are mixed sulphuric and nitric esters, nitrous esters, nitric esters of sugars, and substances produced by the action of the mixed acids on seed husks and on the dried juices in the tubular cotton fibres. These are eliminated or rendered harmless by prolonged boiling of the gun-cotton in large wooden vats (Fig. 33), the water being changed at intervals.

Dr Robertson (*Journ. Soc. Chem. Ind.*, 16th July 1906) has shown the advantage of acidity in the earlier boilings for promoting rapid hydrolysis of the mixed sulphuric and nitric esters.

Treatment with alkaline solutions is not so efficient and may produce hydrolysis of the gun-cotton itself to a considerable extent. The method of boiling in use at Waltham Abbey for periods of 12 + 12 + 4 + 4 + 4 + 4 + 4 + 2 + 2 + 2 hours, with a cold water washing between the first and second and second and third boilings, produces a stable gun-cotton. The nature of the water supply affects the period necessary for stabilisation, and a system in which the gun-cotton is wrung in centrifugals between the boilings, or one in which the pulped gun-cotton is subjected to boiling water treatment, will shorten the time necessary for complete purification. Boiling in closed vessels under pressure has been tried with good results.

The following table is taken from Sir Frederic Nathan's paper :—

Process.	Nature of Dipping Vessel.	Acids.					Cotton Waste Used, lbs.	Acid Used per lb. Cotton Waste.	Time of Nitra- tion. Hours.	Yield on Dry Cotton Waste per Cent.	Output per Man per Week.
		Analysis per Cent.				Quan- tity, lbs.					
		H_2SO_4.	HNO_3.	HNO_2.	H_2O.						
Abel - -	Cast-iron pan and earthen- ware pot	74.00	18.00	0.60	7.40	13.75	1¼	11.0	12	163.75	458
Ardeer—direct dipping	Cast-iron pot	75.00	15.75	1.30	7.95	127	4½	28.2	12	159.00	1,112
Dartford—nitrat- ing centrifugal	Centrifugal machine	69.35	23.15	...	7.50	800 to 1,100	16 to 24	50.0	1	160.00	...
Waltham Abbey —displacement	Earthenware pan	70.50	21.00	0.60	7.90	650	20	32.5	2½	170	1,742

The gun-cotton is then pulped in a beating engine similar to those used for pulping the raw material in paper manufacture (see Fig. 34). The fibres are thus materially shortened and crushed, and impurities retained in the fibres are removed mechanically or by diffusion. The gun-cotton pulp is then freed from particles of metal, grit, and foreign bodies by running it, suspended in a large volume of water, along a shallow trough having grit traps placed in it at intervals. These retain the heavier gritty particles, and fine sand is retained by a strip of blanket which is laid along the bottom of the trough.

At Waltham Abbey the pulp is next washed in a vessel termed a "poacher" (Fig. 34). These hold each about 10 cwt. of gun-cotton and 1,100 gals. of water, and are fitted with power-driven

FIG. 33. —The Boiling House.

FIG. 34.—" Beaters " and " Poachers."

paddles for agitation purposes. The gun-cotton receives at least three washings, and is allowed to settle down after each washing, the washing water being removed by a skimmer. Poaching serves as an opportunity for effecting the final stage of a system of blending which is carried on throughout the manufacture.

The pulp is then moulded by light hydraulic presses into cylinders $5\frac{1}{2}$ in. high and 3 in. in diameter, which enables it to be dried on fixed racks in the drying house without having recourse to movable drying trays or similar arrangements. Less dust is produced than when the gun-cotton is dried in the form of ordinary pulp.

If intended for use in mines, torpedoes, or other demolition work, the gun-cotton is moulded into suitable shapes, and the moulds are then subjected to powerful hydraulic pressure, amounting to about 6 tons per square inch, to produce the finished slabs or primers.

Brown showed that a large quantity of wet gun-cotton can be detonated by interposing between a detonator and the moist gun-cotton a primer of dry gun-cotton (British Patent, 3,115, 1868). Recently a primer of trinitro-toluene has been proposed (D.R. Patent, 200,293, 1906). The pressing of large masses of gun-cotton which are then turned in a lathe to shapes required has been successfully carried out in recent years (Hollings, British Patent, 8,278, 1903).

Claessen (D.R. Patent, 200,262, 1906) proposes to displace the nitrating acids retained in nitro-cellulose by sulphuric acid and the latter by water. Comte de Briailles (D.R. Patent, 203,377, 1906) sends an electric current through the acid mixture during nitration in order to keep the nitrating acid at suitable concentration and temperature. Sir F. L. Nathan, J. M. Thomson, and W. T. Thomson (British Patent, 7,269, 1903) remove the water from wet nitro-cellulose by displacement with alcohol. The Salpetersaure-Ind.-Ges. (D.R. Patent, 180,587, 1906) revivify the waste acid after nitration by electrolysis, using the waste acid as anode liquid, and dilute nitric acid in the kathode compartment.

Assuming the cellulose unit to be $(C_6H_{10}O_5)_4$ or $C_{24}H_{40}O_{20}$, many nitric esters of cellulose are capable of being synthesised, but those met with in practice are comprised among the following (see p. 207) :—

Dodecanitro-cellulose	-	-	$C_{24}H_{28}O_8(NO_3)_{12}$	with 14.14 per cent. nitrogen.	
Endecanitro-cellulose	-	-	$C_{24}H_{29}O_9(NO_3)_{11}$	,, 13.47	,, ,,
Decanitro-cellulose	-	-	$C_{24}H_{30}O_{10}(NO_3)_{10}$	,, 12.75	,, ,,
Enneanitro-cellulose	-	-	$C_{24}H_{31}O_{11}(NO_3)_9$	,, 11.96	,, ,,
Octonitro-cellulose	-	-	$C_{24}H_{32}O_{12}(NO_3)_8$	,, 11.11	,, ,,

The properties of the nitrated product depend on the quality of the cotton waste, the proportion of mixed acid to cotton, the moisture content of the atmosphere, the ratio of H_2SO_4 to HNO_3, the water content of the acid mixture, and on the time, temperature, and method of nitration. Many years ago (1901) Mr W. T. Thomson of Waltham Abbey prepared a specimen of nitro-cellulose having a nitrogen content of 13.94 per cent., obtained by drowning the nitrated product in alcohol, but under ordinary working conditions the gun-cotton contains 12.8-13.10 per cent. of nitrogen. The solubility of the product in ether alcohol, formerly thought to be simply a function of the nitrogen content, is influenced by the method of nitration. Collodion cotton, containing 11-12 per cent. nitrogen, is soluble in ether alcohol. According to Lunge (*Zeitschr. Angew. Chem.*, 2,051, 1906), nitro-cellulose soluble in ether alcohol is obtained by nitrating dry cotton in a mixture of equal parts HNO_3 and H_2SO_4 together with 17-18 per cent. of water, *i.e.*, 41 per cent. H_2SO_4, 41 per cent. HNO_3, 18 per cent. H_2O. Guttmann (*Zeitschr. Angew. Chem.*, 1907, 202) gives an example from practice in which equal parts of nitric acid of 75 per cent. monohydrate and sulphuric acid of 96 per cent. monohydrate (also a mixed acid with 14.5 per cent. water) are used in 25 parts of which 1 part of cotton is nitrated for 1-1½ hours at 40° C.

Although by strict adherence to a definite set of conditions during manufacture a product of unvarying properties is obtained, this is never an individual cellulose ester but a mixture of esters in which one or two predominate. Berl (*D.R. Patent*, 199,885, and *Zeitschr. Sch. Sprengstoffwesen*, 81, 1909) proposes to depolymerise the cellulose molecule by heating the cotton in inert gases. Nitro-celluloses prepared from depolymerised cellulose are claimed to be better than those made from ordinary cotton as regards solubility and capacity for gelatinisation. Collodion cotton finds extended use in the manufacture of gelatinised nitro-glycerine explosives. It is also used in photography and surgery and in large quantities for the manufacture of celluloid and artificial silk. The total world's production of artificial silk is estimated at 5.5 million kilos., of which 2 million kilos. are collodion silk (Massot, *Zeitschr. Angew. Chem.*, 433,444, 1911), see pp. 207, 208, 212.

Of other nitric esters, those from sugar, mannite, and starch are the most important, but various defects have prevented the adoption of the first two to any considerable extent. **Nitro-starch**, on the other hand, is now used quite largely in America for blasting explosives, and its cost of production is said to be considerably less than that of nitro-glycerine. Hough (British Patent, 12,627, 1904) dissolved starch in nitric acid at a temperature of 90° F., and precipitated the nitro-starch by passing gaseous sulphuric anhydride into the solution. In a later patent he altered his process by nitrating the starch with a mixture of 3 parts of nitric acid of 95 per cent. monohydrate and 2 parts of 98 per cent. sulphuric acid, and adding so much sulphuric anhydride to it as to furnish a concentration of 100 per cent. with 1 to 2 per cent. of SO_3 in the solution. Further, during nitration he injects more sulphuric acid having an excess of about 2 per cent. anhydride, and claims to obtain in this way an octonitrate having the formula $C_{12}H_{12}(NO_2)_8O_{10}$ and a nitrogen content of 16.5 per cent. The product is said to be quite stable and to be suitable for use as a blasting explosive either alone or when it forms 10 per cent. of a mixture containing sodium nitrate and carbonaceous material. It has been utilised in the manufacture of smokeless powder. Berl and Bütler (*Zeitschr. Sch. Spr.*, 82, 1910) have tested Hough's data and have found them incorrect. The maximum percentage of nitrogen in nitrate of starch was found by them to be 13.44 per cent.

Dynamite, Blasting Gelatine, Gelatine Dynamite, and Gelignite

With the object of rendering nitro-glycerine safe for transport and use, Alfred Nobel devised the mixture with kieselguhr known as dynamite. Nobel's guhr dynamite consisted of 75 per cent. of nitro-glycerine and 25 per cent. of kieselguhr, an earth containing tubular siliceous skeletons of diatoms which performed the function of a physical container or carrier. Various inert substances have been used instead of kieselguhr, *e.g.*, magnesia alba, tripoli, and Boghead coal ashes. Among dynamites with combustible absorbents may be mentioned Reid and Borland's carbo-dynamite in which burnt cork charcoal is used. It contains 90 per cent. of nitro-glycerine and is consequently more powerful than dynamite, the more so as the absorbent is combustible. Another class of dynamite contains absorbents which are themselves explosive, and a further subdivision of this class may be made into those which contain nitro-cellulose and those which do not.

Kieselguhr dynamite formed only 0.4 per cent. of the total amount of explosives used in mines and quarries during 1909 (H.M.I. report).

For its manufacture the guhr is freed from moisture and organic matter by ignition and is then crushed and sifted. A test is made to determine its capacity for absorbing nitro-glycerine, and if it should prove to yield a dry crumbling dynamite which would give unsatisfactory cartridges it is suitably blended with another batch possessing less absorptive capacity or is mixed with a small quantity of barium sulphate (say 1 per cent.). On the other hand the dynamite should not contain excess of nitro-glycerine or the latter will exude.

The guhr, containing less than half a per cent. of moisture, is weighed out in equal quantities into a series of bags or cases which are taken on trucks to the

7

purified nitro-glycerine store-house where the required amount of nitro-glycerine is accurately measured and added to each batch of kieselguhr. The bags or cases are then removed to the mixing house where their contents are first well mixed by hand and afterwards passed through a sieve. The dynamite is now transferred to the cartridge huts in bags or wooden boxes lined with sheet zinc, and there is made into cartridges of diameter and length required by a "continuous" type cartridge press. The cartridges, wrapped in acid-free parchmentised or paraffin-waxed papers, are collected from the huts at frequent intervals and taken to the packing house, whence samples are despatched to the laboratory for subjection to the heat test and exudation test.

Kieselguhr dynamite is orange-yellow or reddish brown in colour (sometimes ochre is added to the guhr) and is plastic to the same degree as fresh mould. Its specific gravity is 1.5 when it contains 75 per cent. of nitro-glycerine. Small quantities of it burn quietly when ignited but larger quantities soon explode ; when saturated with paraffin oil it may be ignited without danger in small quantities at a time. A blow of 0.75 kg.-m. will cause its detonation, and it can be exploded by a strong electric discharge.

Direct contact with water brings about separation of the nitro-glycerine, and this constitutes a source of danger when the explosive is used in wet places. Its liability to freeze is another source of danger. Very strong detonators are required to explode it when frozen, and though in this condition it is fairly insensitive to a blow, breaking or crushing are distinctly dangerous operations. Thawing must be conducted with great care in a specially constructed pan provided with a jacket for hot water. Should exudation of nitro-glycerine take place during thawing this will introduce an element of danger in ramming.

During storage, exposure to the influences of damp and sunlight must be avoided. About 1 per cent. of sodium or calcium carbonate is frequently added to the ordinary ingredients in order to preserve neutrality.

Rhenish dynamite contains 70 per cent. of a solution of 2-3 per cent. of naphthalene in nitro-glycerine, 3 per cent. of chalk, 7 per cent. barium sulphate, and 20 per cent. of kieselguhr. Nobel's Ardeer powder contains 31-34 per cent. nitro-glycerine, 11-14 per cent. kieselguhr, 47-51 per cent. magnesium sulphate, 4-6 per cent. potassium nitrate with $\frac{1}{2}$ per cent. of ammonium or calcium carbonate.

Dynamites with a combustible base include carbo-dynamite and others in which the absorbent is wood-meal or sawdust. These latter may be detonated when wet and can be dried before use without marked alteration.

The following are examples of mixed dynamites containing an explosive base in addition to nitro-glycerine : 40 per cent. nitro-glycerine, 12 per cent. wood-meal, 46 per cent. sodium nitrate, and $\frac{1}{2}$-1 per cent. calcium or magnesium carbonate. In a French ammon-dynamite, ammonium nitrate is used. It has the following composition : 40 per cent. nitro-glycerine, 10 per cent. wood-meal, 10 per cent. sodium nitrate, 40 per cent. ammonium nitrate.

In mixed dynamites the nitro-glycerine is capable of being displaced by water, as in the case of guhr dynamite. This potential cause of accidents is removed by transforming the nitro-glycerine into a component of a colloidal solution, which is effected by dissolving 7-10 per cent. of collodion cotton in 93-90 per cent. of nitro-glycerine whereby **Blasting gelatine,** a tough elastic mass and one of the most powerful blasting explosives, is formed.

A colloidal solution containing 96-97 per cent. of nitro-glycerine and 4-3 per cent. of collodion cotton, on being mixed with saltpetre and wood-meal in a kneading-machine, yields **Gelatine dynamite,** or **Gelignite,** of which the following is a typical example : 62.5 per cent. nitro-glycerine, 2.5 per cent. collodion cotton, 25.5 per cent. sodium or potassium nitrate, 8.7 per cent. wood-meal, and 0.8 per cent. soda. Another containing ammonium nitrate has the following composition : 50 per cent. nitro-glycerine, 2.5 per cent. collodion cotton, 45 per cent. ammonium nitrate, 2.5 per cent. rye-meal, and is known as **ammongelatindynamite.**

Gelignite formed 10.2 per cent. of the total quantity of explosives used in mines and quarries in this country during 1909, and blasting gelatine and gelatine dynamite together formed 2 per cent. (H.M.I. report)

For the manufacture of these explosives finely pulped and dried collodion cotton is chosen containing not less than 11 per cent. of nitrogen, and having a maximum solubility in ether alcohol and a minimum of gun-cotton or unnitrated cotton. It

FIG. 35. — Messrs Werner, Pfleiderer, & Perkin's
Mixing Machine.

should be free from sand or grit and leave a maximum residue of 0.25 per cent. on ignition. The moisture present should amount to not more than 0.25-0.5 per cent., so exposure to the air is avoided during transit from the drying house to the sifting house. Here the cotton is rubbed gently through a brass sieve, and after being weighed out into air-tight zinc-lined cases or india-rubber bags is removed to the gelatine mixing house where the necessary quantity of pure dry nitro-glycerine is poured upon it in a lead-lined trough. The mixture is well stirred with a wooden paddle and kept for half an hour at a temperature of 40° C. by means of hot-water circulation through the jacket with which the tank is provided. Arrangements should be made for the admission of cold water to the jacket in case the mixture becomes too hot. When the jelly has become semi-transparent it is transferred to the mixing machine, of which an illustration is shown (Fig. 35). M'Roberts' mixing machine is still in use, however (Fig. 36).

At this point, if gelignite or gelatine dynamite is being made, the necessary proportions of purified wood-meal and potassium nitrate are added, together with a certain amount of one of the substances proposed for depressing the freezing point of nitro-glycerine and a little magnesium carbonate. During mixing the charge is maintained at a temperature of 40° C. by

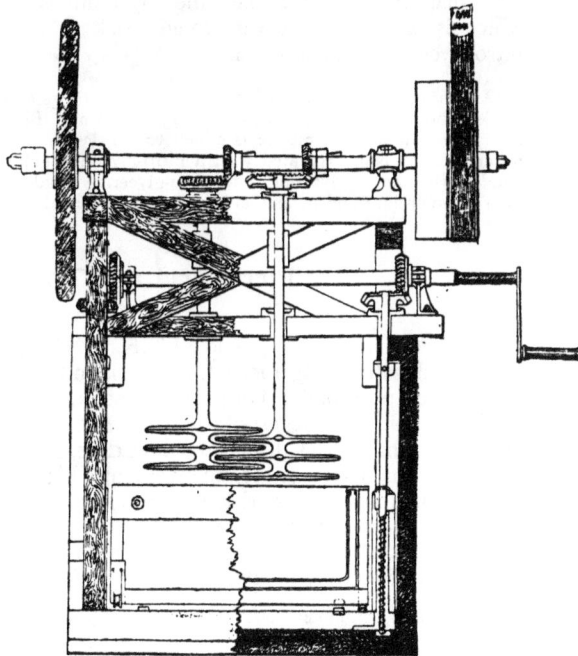

FIG. 36. — M'Roberts' Mixing Machine.

means of the hot-water jacket surrounding the mixing trough.   Incorporation requires at least half an hour.

The paste thus obtained is transferred from the mixing house to the cartridge huts, in which it is made into cartridges by an Archimedean screw machine of the type shown in Fig. **37** ; the cylinder of paste issuing from a nozzle of required diameter is cut into the lengths appropriate for each size of cartridge.

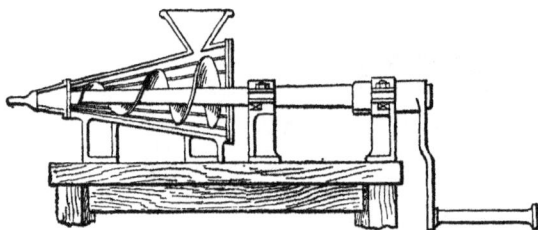

FIG. **37**. —Cartridge Mixing Machine for Gelatine Explosives.

Homogeneity, stability, complete absorption of nitro-glycerine, and freedom from tendency to exude are of the highest importance in these explosives.

Blasting gelatine is an amber-coloured, translucent, elastic mass which stiffens somewhat as time elapses. If well made it resists the conditions of the test for exudation and stands the heat test for a longer period than the minimum of ten minutes.   It explodes at 204° C. if heated slowly (Hess), and at 240° C. if the heating is rapid.   It is also exploded by a blow of $3\frac{1}{2}$ kg.-m. between steel and steel, but is much more sensitive to shock when frozen.

Gelatine dynamite is equally sensitive in its normal soft condition and when frozen, and in both cases is exploded by a weaker blow than is blasting gelatine. The insensitiveness of the latter when it contains camphor is somewhat remarkable. The shattering effect of gelatine dynamite is less than that of blasting gelatine, which is only used for very tough rock or for military purposes, and resembles nitro-glycerine itself in its effects.

Among recent suggestions for the production of gelatinous explosives may be mentioned those of Schachtebeck (British Patent, 22,645, 1902) and Bichel (British Patent, 23,846, 1902).   In the former, ammonium nitrate is added to a gelatine consisting of glue or dextrin, glycerine and water, and this is incorporated with a nitro-glycerine gelatine and a carbohydrate.   In the latter, glue is dissolved in glycerine, and nitro-glycerine is added.   Collodion cotton or a suitable dope such as wood-flour and potassium nitrate may be added before the nitro-glycerine.

Picric Acid and Trinitro-toluene

Picric acid or trinitro-phenol, $\left(C_6H_2 \diagup_{\displaystyle (NO_2)_3}^{\displaystyle OH}\right)$, the oldest organic dye-stuff, has been used extensively for filling shells, and its salts have formed constituents of priming compositions and of powders used as propellants.   The honey-yellow colour of the molten acid suggested the name **Melinite** which it receives in France, while in other countries it is designated **Lyddite, Pertite, Ecrasite**, or **Schimose**. Its melting point, 122.5° C., is inconveniently high, so in order to secure greater ease in manipulation and less risk a quantity of a soluble substance sufficient to depress its melting point considerably is added to it.   Dinitro-toluene, mononitro-naphthalene, and camphor have been used for this purpose.   Girard (British Patent, 6,045, 1905) gives a list of melting points of explosive mixtures of this kind :—

Mixture in Equimolecular Proportions of—				M.P.	M.P. of Mixture.
Trinitro-phenol	-	-	-	122° C.	49° C.
Nitro-naphthalene	-	-	-	61° C.	
Trinitro-phenol	-	-	-	122° C.	47° C.
Dinitro-toluene	-	-	-	71° C.	
Trinitro-phenol	-	-	-	122° C.	70° C.
Trinitro-cresol	-	-	-	107° C.	

Mixture in Equimolecular Proportions of—					M.P.	M.P. of Mixture.
Trinitro-phenol	-	-	-	-	122° C.	
Trinitro-cresol	-	-	-	-	107° C.	78° C.
Trinitro-phenol	-	-	-	-	122° C.	
Trinitro-cresol	-	-	-	-	107° C.	80° C.

The specific gravity of picric acid solidified from fusion is 1.64. A priming composition, such as a mixture of ammonium picrate and potassium nitrate, is necessary to detonate the hard mass.

The manufacture of picric acid is carried out in the following way (*Zeitschr. Schiess. Sprengstoffwesen*, 15, 1910).

Phenol-sulphonic acid is first prepared, $\left(C_6H_4{<}^{OH}_{SO_3H}\right)$, which is then transformed into dinitro-phenol-sulphonic acid $\left(C_6H_2(NO_2)_2{<}^{OH}_{SO_3H}\right)$. The latter decomposes on warming with nitric acid according to the equation :—

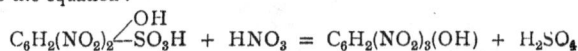

$$C_6H_2(NO_2)_2{<}^{OH}_{SO_3H} + HNO_3 = C_6H_2(NO_2)_3(OH) + H_2SO_4$$

12 parts by weight of phenol (**M.P.** 40° C.) are heated to 100°-105° C. for eight hours with 48 parts of sulphuric acid of sp. gr. 1.84, with constant stirring, in a vessel provided with a steam jacket. The melt is drawn off into a transportable earthenware vessel in which, on cooling, it solidifies to a butter-like mass. This is transferred by means of a scoop, in small portions at a time, to 90 parts of nitric acid of sp. gr. 1.385 contained in a large earthenware vessel which stands in a water-bath, stirring being continuous during the whole operation, and the temperature being kept below 20° C. This causes the phenol-sulphonic acids to change into dinitro-phenol-sulphonic acid. When all the sulphonic acid has been introduced, which for 100 kg. requires about twenty hours, the water surrounding the nitrating vessel is slowly warmed, by leading in steam, so that the temperature rises to 40°-50° C. within the first four hours, and the water is then heated to boiling point. At 80°-90° C., large quantities of nitrous fumes are evolved which are conducted through earthenware pipes to absorption towers; the vigorous reaction having subsided, the water in the heating bath is kept boiling for three to four hours longer, and the contents of the vessel are stirred at intervals with a strong glass or porcelain rod. Heating is then discontinued, and the contents of the vessel are allowed to cool to 30°-35° C., when the picric acid appears almost entirely in the crystalline form. The waste acid is syphoned off, and several water washings are followed by wringing in a centrifugal, which is lined with woollen cloth, and into which cold water is sprayed until the drainings contain traces only of H_2SO_4. After wringing, the picric acid still contains 10-15 per cent. of water, and its further drying is effected by placing it on glass plates in a warm air current or in a vacuum drying apparatus. To recrystallise (not usually necessary) dissolve moist crystals in 10-12 times their weight of boiling water, filter through wool, and slowly cool.

The yield from 100 parts of phenol, melting at 40° C., amounts to not more than 190 parts of picric acid, of melting point 121°-122° C., while theory requires 243 parts. The diminution in yield is to be attributed to oxidation due to the excess of nitric acid used. At the same time a considerable excess of nitric acid is necessary in order to avoid the production of lower nitro-derivatives of phenol.

Picric acid can also be made by nitrating phenol with **sodium nitrate and sulphuric acid.** [See German Patents, 51,321 (1889), and 67,074 (1891).]

The high melting point of picric acid is not its only disadvantage. It will have been noticed that during manufacture all contact with metals is avoided. Some of the metallic picrates are extraordinarily sensitive to shock, for instance those of lead and copper, and these salts are formed if the acid comes in contact with metals (except tin) or their oxides. Under favourable circumstances their detonation may take place, and, simultaneously with it, that of all the picric acid surrounding them, or in the immediate neighbourhood, hence the necessity for varnishing the interiors of shells, for giving special protection to primers, and for taking the utmost precaution to prevent access of foreign bodies. Picric acid will generate other acids from salts containing their radicles, and will decompose nitric esters, and if used at all it should be used alone. The dust from it is poisonous, as also are the fumes arising from the molten acid, and its dyeing effect upon the skin is well known. **Ammonium picrate** and **nitre**, the constituents of **Brugère's** powder, are now used as a priming composition. **Solubility of Picric Acid.**—100 parts water dissolve 0.626 at 5° C., 1.161 at 15° C., 1.225 at 20° C., 1.380 at 26° C., 3.89 at 77° C. H_2SO_4 diminishes its solubility. Easily soluble in alcohol and ether; benzol dissolves 8-10 per cent. at ordinary temperatures.

Trinitro-toluene has sprung into prominence as a substitute for picric acid, chiefly owing to its freedom from acidic properties, its lower melting point, and great stability. Of the various isomers the one having the structure

$C_6H_2(NO_2)_3.CH_3 I.(_{CH_3}: 2:4:6)$ is used (see *Ber.*, 1914, p. 707; *Zeitsch. gesammte Schiess. Sprengstoffwesen*, **1907**, 4; **1910**, 156; **1911**, 301; **1912**, 426; **1914**, 172). Its manufacture from pure toluene is carried out in stages.

180 kg. of pure toluene are first of all transformed into **mono-nitro-toluene** by means of a mixture of 315 kg. of sulphuric acid (sp. gr. 1.84) and 200 kg. of nitric acid (sp. gr. 1.44); the waste acid is drawn off, and a fresh acid mixture of 600 kg. of sulphuric acid (sp. gr. 1.84) and 200 kg. of nitric acid (sp. gr. 1.48) transforms the first product into dinitro-toluene. The mixture is heated and stirred during the operation, and the dinitro-toluene separates as an oily layer from the hot mixed acids. The waste acid from this second stage is revivified by the addition of 135 kg. of fresh nitric acid (sp. gr. 1.44), and is used again for making mono-nitro-toluene.

Dinitro-toluene forms, on cooling, a pale yellow, fibro-crystalline, moderately hard and brittle mass, which on further treatment with sulphuric and nitric acids is transformed into trinitro-toluene. For this purpose it is dissolved by gently heating it with four times its weight of sulphuric acid (95-96 per cent.) and it is then mixed with one and a half times its weight of nitric acid (90-92 per cent.), the mixture being kept cool. Afterwards it is digested at 90°-95° C., with occasional stirring, until the evolution of gas ceases, which takes place in about four or five hours. The product, after cooling, is separated and washed.

According to the purity of the **dinitro-toluene** one may obtain an end-product with a melting point of 72°-73° C., or a purer form with a solidifying point of 77°-79° C. The impure form is used exclusively in blasting explosive mixtures. By recrystallising the higher melting product from alcohol or petroleum benzene it may be obtained pure, and then melts at 80.6° C. It is finely crystalline, is bright yellow in colour, has no specially bitter taste, and the dust arising from it is said to have no injurious effect. Whether in the form of powder, or in the massive condition after pressing or fusion, it is quite insensitive and much safer to manipulate than picric acid. It can be detonated with a No. 3 detonator (0.54 g. of fulminate composition) when in the form of powder, and no poisonous gases are produced. It is quite stable under the influence of heat and moisture, and has no action upon metals or metallic oxides.

Its low density is a drawback (1.54 against 1.64 for picric acid). Rudeloff (*Zeitschr. Schiess. Sprengstoffwesen*, 7, 1907), obtains a density of 1.85 to 1.90 by making a plastic substance from trinitro-toluene and potassium chlorate with a gelatine made from dinitro-toluene and soluble nitro-cellulose. Bichel (British Patent, 16,882, 1906) makes a plastic "compound" with collodion cotton, liquid dinitro-toluene, aud larch turpentine to which he gives the name **Plastrotyl**. Messrs Allendorf mix the trinitro-toluene, together with some lead nitrate or chlorate, with a gelatine made from dinitro-toluene and nitro-cellulose, and call the product **Triplastit**. Bichel (British Patent, 19,215, 1906) melts the trinitro-toluene and, after first exhausting all occluded air, compresses the molten product during cooling by means of compressed air. In this way he increased the density to 1.69. Rudeloff, using hydraulic presses, subjects it to a pressure of 2,000-3,000 atmospheres whereby it attains to a density of 1.7, and can be cut and worked like gun-cotton. For the purpose of facilitating detonation some loose trinitro-toluene is used as a primer.

Trinitro-toluene has been introduced into the French service under the name of **Tolite**. The Spanish Government call it **Trilit**. The carbonite works of Schlebusch are introducing it into other services under the name of **Trotyl**, and Messrs Allendorf of Schoenebeck under the name of **Trinol**.

Trinitro-benzene is also receiving attention as a suitable material for filling shells.

Schroetter and von Freiherr (British Patent, 8,156, 1907) have proposed the use of **hexa-nitro-diphenylamine**, in a compressed condition, or its salts, in the manufacture of high explosives for adjusting the sensitiveness and the melting point.

Other nitro-derivatives of hydrocarbons and of phenols are used in explosive mixtures. Among the former are **dinitro-benzene, chlor-dinitro-benzene, dinitro-toluene, di-** or **tri-nitro-derivatives** of **mesitylene, pseudocumene,** and **xylene, dinitro-naphthalene. Trinitro-naphthalene** and **tetra-nitro-naphthalene** have also been proposed. Of the nitro-phenols, perhaps the most important is **tri-nitro-cresol**, known as **Cresilite** in France, where it is added to **Melinite.** Other nitro-derivatives of the aromatic series will be mentioned in the section on detonators.

The work of Dr B. J. Flürsheim on the polynitro-derivatives of aromatic compounds has aroused widespread interest. 2:3:4:6-**tetranitro-aniline** or **tetranyl**, which he obtains by nitrating 2:3-dinitro-aniline, 3:4-dinitro-aniline or metanitro-aniline (British Patents, 3,224, 1910, and 3,907, 1910; D.R.P., 242,079 and 241,697), appears from results of tests which have been published to be particularly well adapted for use as an explosive (*Proc. Chem. Soc.*, 26, 1910; *Zeitschr. Schiess. Sprengstoffwesen*, 10, 1913, and 2, 1915; *Journ. U.S. Artillery*, 40, 1913), and is suggested for application in detonators, explosive mixtures, propulsive powders, and detonating fuses, also for filling shells, mines, and torpedoes. It may be used as a primer (British Patents, 16,673, 1911, and 2,407, 1912) or in admixture with (1) **trinitro-amino-anisol** or **trinitro-amino-phenetol** (British Patent, 18,777, 1911) or (2) with aluminium and ammonium perchlorate (French Patent, 465,082, 1913). The application of tetranitro-aniline on an appreciable scale will no doubt depend on whether or not the stringent official requirements regarding explosives and their ingredients can be met at a suitable cost and without sacrifice of efficiency.

Ammonium Nitrate Explosives, Safety Explosives

Bichel, in conjunction with Dr Mettegang and other colleagues, has investigated experimentally the conditions which determine the superiority of some explosives over others, with regard to their liability to ignite mixtures of pit gas and air. They have proved that this liability depends upon the length, duration, and temperature of the flame accompanying detonation, and the figures obtained on dividing the duration of flame by the time of detonation, to which they give the name "**after-flame**" ratio, distinguish in no uncertain fashion those explosives which may be considered "**safe**" from those which are known to be **unsafe**. Even a "safe" explosive becomes dangerous if more than a certain amount of it, known as the "**charge limit**," be used at one time.

In testing explosives for "safety" at the British Home Office Testing Station at Rotherham shots are first fired into a mixture of gas and air (13.4 per cent. coal gas). When the largest charge which can be fired, without igniting the mixture, has been determined, further shots are fired— beginning with this charge, and, in the event of an ignition, reducing the charge—until five shots of the same weight have been fired without igniting the mixture. Starting with the charge so determined similar experiments are carried out with coal dust until five shots of the same weight have been fired without igniting the coal dust. The lower of the charges thus determined is known as the "Maximum Charge." The shots are unstemmed. In Germany a more dilute gas mixture is used (7.9 per cent. methane in air). The less the area of cross section of the gallery the more easily does ignition take place, so until a standard type of gallery is agreed upon, and a definite method of testing adhered to, the results from the various testing stations can have little value for purposes of comparison.

Gunpowder, trinitro-toluene, picric acid, blasting gelatine, gelatine dynamite, and gun-cotton are all unsafe. Explosives containing nitro-glycerine may, however, be perfectly safe; in fact, such explosives are amongst the safest we have. In these the percentage of nitro-glycerine is comparatively low, and the other ingredients are such as will give as low a temperature as is consistent with satisfactory work. Ammonium nitrate is very frequently used in safety explosives, and a small amount of nitro-glycerine will ensure its complete detonation. Many safety explosives contain ammonium nitrate and compounds of the type of dinitro-benzene, while in others, which contain no ammonium nitrate, the cooling effect of substances like ammonium oxalate or salts containing water of crystallisation is relied upon to confer the necessary degree of safety. While ammonium nitrate explosives in general may be said to have a higher "charge limit" and a less violent shattering effect than the explosives mentioned above, and possess the further advantages of being safer to handle and transport and of requiring a strong detonator, it must not be forgotten that their use is best confined to a definite type of work. The hygroscopic nature of ammonium nitrate is a disadvantage which is overcome by various devices, but its low specific gravity necessitates greater expense in preparing the larger bore-holes necessary to accommodate the charge.

For use in the production of explosives the ammonium nitrate is dried, finely ground, mixed with the other ingredients in a drum-mixer, again dried at 60° C., and afterwards made into cartridges. The cartridges for protection against moisture are either surrounded by tinfoil or impregnated with paraffin or ceresine, which is accomplished by dipping them in a bath containing these substances in the molten condition.

Rigid cartridges are produced by pressing a mixture made plastic either by means of a molten binding material (a nitro-hydrocarbon) or by a solution—collodion cotton in acetone for example. On cooling, or on evaporation of the solvent, the solid cartridge is obtained.

The following are explosives containing ammonium nitrate: Ammoncarbonite, Wetterfulmenite, Chromammonite, Roburite, Dahmenite, Dorfite, Bellite, Securite, Trench's flameless explosive, Faversham powder, Ammonite, Electronite, Amvis, etc.

As mentioned above, the addition of various salts to the explosive mixture for the purpose of reducing the temperature of the detonation products has been made. Claessen (British Patent, 2,240, 1906) uses chrome alum or ammonium chrome alum; the Castroper Sicherheitssprengstoff Akt.-Ges. (British Patent, 18,275, 1905) use ammonium chloride and potassium oxalate; Curtis and others (British Patent, 24,934, 1902) suggest copper sulphate, or a mixture of copper sulphate with ammonium sulphate or equivalent salts. The fire extinguishing compound of Mr Geo. Trench, consisting of sawdust impregnated with a mixture of alum and chlorides of sodium and ammonia, is so disposed about the explosive proper as to prevent the ignition of a surrounding explosive gas mixture. Nearly all safety explosives require a No. 6 detonator and are generally fired by low tension fuses connected to a magneto firing apparatus. The composition of a few members of this large class is given in the appendix to this article. An apparently practical proposal for supervision of the mine gases is made by Teclu (*J. Prakt. Chem.*, 82, 237-240, 1911, Chem. Lab. der Wiener Handelsakademie) who tests the gas mixtures at various parts of the mine by drawing samples through tinned iron tubes, by means of a powerful suction pump, into explosion vessels situated on the surface. The gas samples can be tested in a few minutes with perfect safety, and the results signalled.

Plasticity is a desirable quality in an explosive intended for use in bore-holes, and ammonium nitrate explosives possessing plasticity are now produced.

Gelatineastralite contains ammonium nitrate with some sodium nitrate, up to 20 per cent. of dinitro-chlorhydrin and maximum amounts of 5 per cent. of nitro-glycerine and 1 per cent. of collodion cotton. **Gelatinewetterastralite 1** has the composition: 40 per cent. ammonium nitrate, 7.5 per cent. sodium nitrate, 16 per cent. dinitro-chlorhydrin, 4 per cent. nitro-glycerine, 0.5 per cent. collodion cotton, 0.5 per cent. wood-meal, 8 per cent. potato starch, 2 per cent. rape seed oil, 1 per cent. nitro-toluene, 2 per cent. dinitro-toluene, 14 per cent. salt, and 2.5 per cent. ammonium oxalate. Another plastic ammonium nitrate explosive known as **Plastammone** contains ammonium nitrate, glycerine, mono-nitrotoluene, and nitro-semicellulose.

The use of aluminium powder as an ingredient of explosives, first suggested by Dr Richard Escales in 1899, results in a rise of temperature and a considerable increase in pressure produced by the heat evolved during the oxidation of the aluminium, in other words, the explosive power is increased.

The high price of aluminium powder and its liability to oxidise prematurely have militated against its more general adoption. **Ammonal** is an explosive containing aluminium powder, and a similar mixture has been used in Austria-Hungary for several years for filling shells. The latter has approximately the following composition: 47 per cent. ammonium nitrate, 1 per cent. charcoal, 30 per cent. trinitro-toluene, and 22 per cent. aluminium. It is compressed, heated to 67° C., dipped in molten trinitro-toluene, and cooled in a stream of air.

Calcium-silicon, ferro-silicon, silicon, silicon carbide, magnesium, zinc and its alloys, copper, and the rare metals have all been the subjects of patents regarding the production of effects similar to those of aluminium.

Chlorate and Perchlorate Explosives.

The developments of electro-chemistry having resulted in the production of chlorates in large quantity, an outlet for these was sought in the sphere of explosives and ultimately found. Chlorates have long been forbidden for explosives manufacture on account of the sensitiveness of mixtures containing them to shock and friction. Street found, however, that by using castor oil in the mixture this danger was obviated (D.R. Patent 100,522, 1897; 100,532, 1897; 117,051, 1898; 118,102, 1898). **Cheddites**, so-called from their place of manufacture, Chedde, in Haute-Savoie, are composed as follows (Dr Escales' article on " Explosivstoffe," Dammer's " Chemische Technologie der Neuzeit "): (1) $KClO_3$ 79 per cent., nitro-naphthalene 15 per cent., castor oil 6 per cent.; (2) $KClO_3$ 79 per cent., dinitro-toluene 15 per cent., nitro-naphthalene 1 per cent., castor oil 5 per cent; (3) $NaClO_3$ 75 per cent., dinitro-toluene 19 per cent., nitro-naphthalene 1 per cent., castor oil 5 per cent.; (4) $KClO_3$ 80 per cent., dinitro-toluene 2 per cent., nitro-naphthalene 10 per cent., castor oil 8 per cent.

The dinitro-toluene and nitro-naphthalene are dissolved in warm (60°-70°) castor oil contained in a double-walled enamelled iron vessel, and the finely pulverised, dried and sieved potassium chlorate is added with constant stirring or kneading by means of a large wooden paddle. After further working the explosive is corned and on standing for some time is made into cartridges which are waterproofed.

A **cheddite** containing ammonium perchlorate has the following composition: 82 per cent. ammonium perchlorate, 13 per cent. dinitro-toluene, and 5 per cent. castor oil. HCl is evolved in its decomposition, and is not entirely prevented from forming by the addition of sodium nitrate. It possesses, in a high degree, the disadvantage, shared by chlorate cheddites, that it easily becomes hard, and is then not completely detonated.

Colliery steelite consists of 74 parts of potassium chlorate, 25 parts of oxidised resin, and 1 part of castor oil.

Silesia Powder (186,829, 1902, D.R. Patent) is a mixture of 75 per cent. potassium chlorate and pure or nitrated resin. **Permonite** and **Alkalsite** contain 25-32 per cent. potassium perchlorate, ammonium nitrate, trinitro-toluene, and other constituents. **Yonckite**, a Belgian explosive, contains, in addition to perchlorate, potassium nitrate and mononitronaphthalene.

Spermaceti,[*] tar,[†] naphthalene [‡] and paraffin, linseed oil,[§] and a solution of hydrocarbons [||] have all been proposed for use with chlorate explosives.

The suggestion, made by Sprengel so long ago as 1871, to transport an oxygen carrier and a combustible substance separately to the place required, and there to mix them in suitable proportions immediately before use, is still acted upon in some countries. Potassium chlorate is one of the oxygen carriers used.

Winand (British Patent, 26,261, 1907) has suggested **tetranitro-methane** as an oxygen carrier for the production of explosives of the Sprengel type.

This substance should prove useful in other respects, though its volatility is a disadvantage. It can be made quite easily, without danger, and in almost theoretical amount by allowing nitric acid to react with acetic acid anhydride at ordinary temperatures (Chattaway, *Journ. Chem. Soc.*, Nov. 1910).

It is a colourless oil which solidifies to a crystalline mass, melting at 13° C. It is insoluble in water, but dissolves readily in alcohol and ether, is very stable, and does not explode on the application of heat, but distils quietly at 126° C.

Fulminate of Mercury. Azides. Primers. Percussion Caps. Detonators

The detonation of an explosive is almost always effected by the detonation of a more sensitive explosive, of which a small quantity is placed in juxtaposition with the first, and fired by mechanical shock, fuses, or electrical devices. Ordinary percussion cap composition contains a mixture of mercury fulminate, potassium chlorate and antimony sulphide to which powdered glass may be added, in order to obtain increase of sensitiveness. The caps for detonators are of pure copper, are cylindrical in shape, closed at one end, and charged with an intimate mixture of 85 per cent. mercury fulminate and 15 per cent. potassium chlorate. Detonators are made in ten sizes, numbered consecutively, and contain 0.3, 0.4, 0.54, 0.65, 0.8, 1.0, 1.5, 2.0, 2.5, or 3 g. of the detonating mixture.

Fulminic acid is the oxime of carbon monoxide, $H.O.N{=}C''$, and its mercury salt is represented by the formula $Hg {<} \begin{matrix} O.N{=}C'' \\ O.N{=}C'' \end{matrix}$. The latter is obtained by the action of ethyl alcohol or of methylated spirit on a solution of mercury in nitric acid. The reaction takes place in a large flask or carboy, and may require to be moderated by the addition of alcohol. It has been shown that oxides of nitrogen must be present; that acetaldehyde gives a much larger yield of fulminate than alcohol; and that 1-carbon bodies (methyl alcohol, formaldehyde) and also 3- and 4-carbon compounds give none at all.

The alcohol is first oxidised to aldehyde: this is converted by the nitrous acid present into isonitroso-acetaldehyde, and this is further oxidised to isonitroso-acetic acid. This last body is converted by nitrous fumes (by nitration followed by loss of carbon dioxide) into methyl nitrolic acid, which reacts with mercuric nitrate to give mercury fulminate :—

$$CH_3.CH_2.OH \rightarrow CH_3.CHO \rightarrow HO.N{:}CH.CHO \rightarrow HO.N{:}CH.COOH \rightarrow HO.N{:}C(NO_2).COOH$$
$$\rightarrow HO.N{:}C(NO_2)H \rightarrow HO.N{:}C$$

Small grey needles of the fulminate separate, and after transference to a cloth filter are washed till acid free, and then dried in a vacuum apparatus. From 100 g. of mercury 125 g. of fulminate are obtained. The alcohol in the condensed vapours is used again. A pure white fulminate, free from metallic mercury, is obtained by adding a little hydrochloric acid and copper to the acid mixture. It is soluble in 130 times its weight of boiling water, and in ammonia, has a sp. gr. = 4.42, and when dry explodes by a moderate blow or slight friction. When detonated in a space entirely filled by it, it develops a pressure more than twice

---

[*] Hahn (British Patent, 960, 1867).   [†] Tschirner (*ibid.*, 447, 1880).
[‡] Fraenkel (*ibid.*, 13,789, 1888).   [§] Brank (*ibid.*, 5,027, 1891).
[||] Hinly (*ibid.*, 1,969, 1882), and Lyte and Lewall (*ibid.*, 14,379, 1884).

that of nitro-glycerine treated under similar circumstances. Confinement is necessary to develop its full power, though a very slight degree of confinement is sufficient.

In the firing of smokeless powders the necessary increase in temperature is obtained by adding a combustible substance. Aluminium powder, either mixed with the fulminate or pressed in a layer on top of it, has been successfully employed by Dr Brownsdon and the King's Norton Metal Company (British Patent, 23,366, 1904). An important step in advance was made when the expensive and dangerous fulminate of mercury or its mixture with potassium chlorate was diminished in quantity to a fraction of its original amount by the use of picric acid or, much better, trinitro-toluene.

A No. 8 detonator containing 2 g. of the ordinary fulminating composition is equalled by one containing only 0.5 g. together with 0.7 g. of trinitro-toluene.

Dr Claessen (British Patent, 13,340, 1905) has shown that even better effects are obtained by the use of what he calls tetranitro-methylaniline or tetranitro-ethylaniline. These substances are really trinitro-phenyl-methyl-nitramine and trinitro-phenyl-ethyl-nitramine. The commercial name for Claessen's substance is "**tetryl**." "Tetryl" detonator No. 6 contains 0.4 g. of trinitro-phenyl-methyl-nitramine and 0.3 g. of a mixture of 87.5 per cent. mercury fulminate and 12.5 per cent. potassium chlorate.

Mercury fulminate may now be dispensed with and replaced by an even smaller quantity of an azide. Dr Lothar Wöhler finds that an ordinary No. 8 detonator is equalled by one containing 1 g. of picric acid or of trinitro-toluene and 0.023 g. of silver azide in place of the usual mercury fulminate charge of 2 g. (British Patent, 4,468, 1908).

M. Hyronimus (British Patent, 1,819, 1908) has patented the use of lead azide, $Pb(N_3)_2$, and describes a safe method of manufacturing it, founded on the process used by Wislicenus (*Ber.*, 1892, 25, 2084).

Sodamide is first prepared, and is transformed into sodium azide by warming at 300° C. in a stream of nitrous oxide gas. The sodium azide is dissolved in water, the solution is neutralised with very dilute nitric acid, and lead azide is precipitated by adding a solution of lead nitrate. The reactions which take place are represented by the following equations : (1) $Na + NH_3 = NH_2Na + H$; (2) $NH_2Na + N_2O = NaN_3 + H_2O$. The water formed decomposes part of the sodamide, thus : (3) $NH_2Na + H_2O = NH_3 + NaOH$. The reaction of the sodium azide with lead nitrate goes as follows : (4) $2NaN_3 + Pb(NO_3)_2 = Pb(N_3)_2 + 2NaNO_3$. The crystalline lead azide is washed on a filter and dried at a temperature below 100° C.

At 300° C. sodamide is fusible and sodium azide infusible ; the latter as it forms absorbs the former and protects it from further action of the nitrous oxide. The sodamide is therefore incorporated with anhydrous materials like lime, magnesia, or sodium sulphate at 190°-200° C., and in this subdivided state reacts more readily with the nitrous oxide. Siliceous absorbents such as pumice stone and infusorial earth would react with the caustic soda formed in the reaction, and both delay the process and diminish the yield.

Thiele (*Ber.*, 1908, 41, 2681-2683) has obtained an almost quantitative yield of azoimide by the action of ethyl nitrite on hydrazine in presence of alkali. Hydrazine hydrate (1 molecule) is mixed with 4 N. sodium hydroxide ($1\frac{1}{2}$ molecules), ethyl nitrite ($1\frac{1}{2}$ molecules), and ether, and the mixture is allowed to remain first in ice and then at the ordinary temperature. After twenty-four hours the sodium azide is collected and washed. Hydrazine sulphate may be used with appropriate modifications. Stollé (*Ber.*, 1908, 2811-2813) has patented a similar process. The yield of azoimide obtained when a benzene solution of nitrogen chloride is shaken with a solution of hydrazine sulphate, with addition of caustic soda at intervals, is 36 per cent. (Tanatar, *Ber.*, 1899, 32, 1399).

Smokeless Powders

Smokeless powders of the present day contain, as chief constituent, nitro-cellulose alone or mixed with nitro-glycerine or some other nitro compound or inorganic oxidising agent. Attempts to use fibrous gun-cotton as a propellant were attended by failure. Dangerously high pressures were developed in the bore of the gun. It was only when the control of the rate of combustion of the explosive was made possible by gelatinisation that progress could take place.

Roughly speaking, gelatinised nitro-cellulose burns in consecutive layers in such a way that equally thick layers are consumed in equal times. With the same composition of explosive it is possible, by varying the shape and size of the unit pieces of powder in the charge, to regulate the rate of combustion of a powder so as to give the desired velocity and yet keep within the pressure limits permissible. In France, powders take the form of ribbons, in Great Britain cords, in Germany flakes and tubes, in Italy cords of square section, in the United States short multiperforated cylinders. Ballistite is formed into cubes, and sporting powders abroad have in some cases a spiral shape.

For gelatinising nitro-cellulose the solvents acetone or ether alcohol are generally used, the former for gun-cotton and the latter for collodion cotton. The nitro-cellulose is used after drying by the ordinary method in drying houses, or is made anhydrous by soaking it in alcohol—first of all in alcohol which has already been used for this purpose and then in pure alcohol, squeezing the liquid out after each operation in a press. Gelatinisation of the finely powdered dry nitro-cellulose takes place in a mixing machine, after which the mixture is rolled under a pair of heavy rolls into sheets of the required thickness (0.3-0.7 mm.) which are afterwards cut up into squares and dried.

The British service powder, Cordite M.D., is a nitro-cellulose, nitro-glycerine colloid containing 65 per cent. of gun-cotton, 30 per cent. of nitro-glycerine, and 5 per cent. of mineral jelly. The percentage of nitro-glycerine was originally 58 and that of gun-cotton 37, but this produced a considerable erosive effect upon the guns which has been minimised by the adoption of the present modification without materially altering the ballistic results.

It is made as follows: Dry gun-cotton is taken in non-absorbent bags to the nitro-glycerine mixing house where the requisite amount of nitro-glycerine is added. The "paste" now in the bags is taken to a sifting house, where it is first mixed by hand in a leaden trough standing on supports and then rubbed through a sieve situated at the bottom of a bowl-shaped extension into a bag suspended beneath. Removal to the incorporating house next follows. In this is a kneading machine of the well-known Werner, Pfleiderer, & Perkins type (Fig. 245), is started after some acetone has been placed in the trough, the "paste" is then added and lastly the remainder of the acetone. After three and a half hours' incorporation the mineral jelly is added and the kneading continued for another three and a half hours. The trough is water-jacketed for cooling purposes, and a cover is placed above it to prevent loss of acetone. The cordite paste is then placed in a cylindrical mould and forced by hydraulic pressure through a die, and the long cord thus formed is wound on a metal drum or cut into lengths. It is then sent to the drying houses and dried at a temperature of about 38° C. for three to fourteen days according to the diameter of the cordite, after which it is thoroughly blended and despatched from the factory.

The finished cordite resembles a cord of gutta-percha, and varies in colour from light to dark brown. Its diameter varies according to the type of gun for which it is intended, and amounts to .0375 of an inch for the .303 rifle when it is used in bundles of sixty strands, while for heavy guns sticks of 0.40-0.50 of an inch diameter and 14 in. long are used. A smaller charge is required to produce a given muzzle velocity with cordite than with gunpowder, and temperature differences do not affect its ballistic properties much more than in the case of the latter. It is safe to handle, is very insensitive to shock, can be burnt away in considerable quantities without explosion, and retains its stability for considerable periods under the most varied climatic conditions. Its erosive effect upon guns is greater than that of gunpowder, and differs in its nature from that produced by the latter, which furrows the metallic surface, while cordite sweeps the surface smoothly away for a shorter distance.

The annexed figure, taken from Professor V. B. Lewes's paper read before the Society of Arts, and due to Dr W. Anderson, F.R.S., shows graphically the pressures given by cordite and by black powder in the 6-in. gun (Fig. 38).

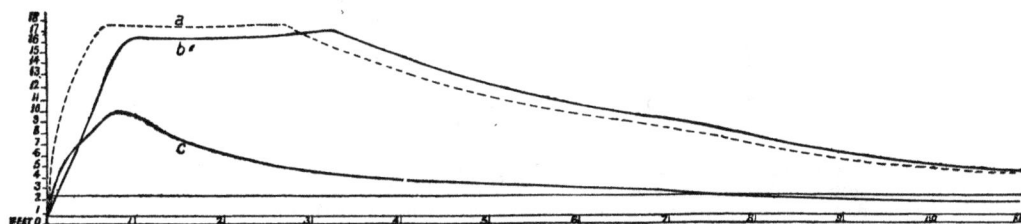

Fig. 38.— Comparative Pressure Curves of Cordite and Black Powder.

a, Charge, 48 lbs. powder; *b,* Charge, 13 lbs. 4 oz. cordite; *c,* Charge, 13 lbs. 4 oz. powder. Weight of projectile, 100 lbs. in 6-in. gun. M.V. cordite = 1,960 ft. per sec.

A problem of great economic importance, that of the profitable recovery of solvent from the gelatinised explosive during drying, has been solved in recent years.   Dr Robert Robertson and Mr Wm. Rintoul (British Patent, 25,993, 1901) have worked out, at Waltham Abbey, the process now used for recovering acetone from the atmosphere of cordite drying houses.

The accompanying drawing of the absorption apparatus (Fig. 39), together with the following brief description, will serve to indicate the method of working.   The interiors of the drying houses are

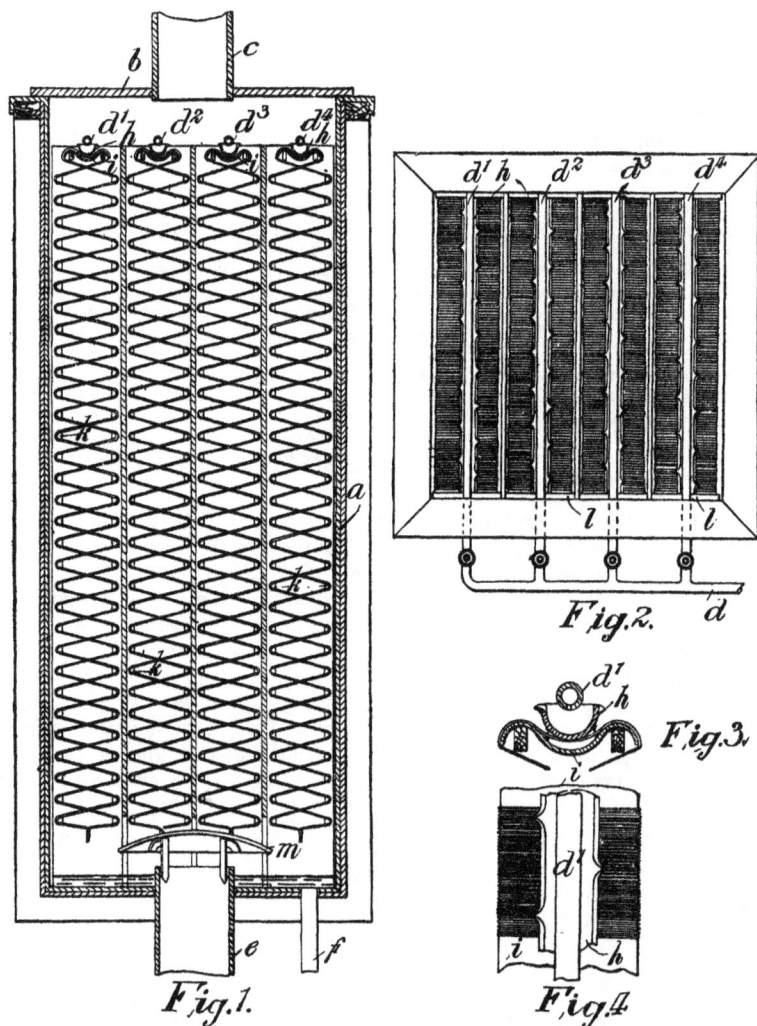

FIG. **39.** —Section, Plan, and details of an Absorption Tower
used in Robertson and Rintoul's Acetone Recovery Process.

connected by zinc pipes to a central absorption apparatus consisting of a series of towers of rectangular section.   A concentrated solution of sodium bisulphite is run down these towers, and is distributed so as to descend by way of the woollen or other fibres which are arranged in zigzag fashion on wooden frames, and at the same time a Root's blower or exhaust fan aspirates the acetone-laden air of the drying houses through the towers in series, the air current passing up the tower while the bisulphite solution flows down.   The liquors which have collected at the bottom of the last tower and have met the poorest acetone mixture are blown by compressed air elevators to the tank connected with the top of the second tower, while fresh bisulphite solution continually descends the last tower. The partially saturated bisulphite solution descends the second tower, and afterwards, in the same way, the remaining towers of the series; the saturated solution obtained from the first tower, that up which the air current first passes, is distilled, and the distillate is rectified over caustic soda. The residual bisulphite solution is used again.

Bouchaud-Praeciq (British Patent, 6,075, 1905) has described a process for alcohol ether recovery in which the vapours are aspirated first through a cooling and drying vessel containing a mixture of calcium carbide and sodium, and thence to an absorbing chamber, charged with pumice stone, down which sulphuric acid flows. The liquid is distilled at a low temperature or under reduced pressure for recovery of the solvent.

Solenite is an Italian nitro-glycerine, nitro-cellulose powder, containing 30 parts of nitro-glycerine, 40 parts of "insoluble" and 30 parts of "soluble" nitro-cellulose, the two latter having an average of 12.6 per cent. of nitrogen. Acetone is used for promoting the solution of the "insoluble" nitro-cellulose.

Ballistite is another nitro-glycerine, nitro-cellulose powder the manufacture of which differs from those already given. "Soluble" nitro-cellulose, in the form of a fine powder, is suspended in fifteen times its own bulk of hot water, and nitro-glycerine is added, the mixture being stirred by means of compressed air (Lundholm & Sayers, British Patent, 10,376, 1889). The water acts as a carrier and enables the nitro-glycerine to gelatinise the nitro-cellulose. The paste resulting is freed from water in a centrifugal machine and allowed to ripen. It is then brought under heated rolls (122°-140° F.), weighted to exert a pressure of 100 atmospheres and the sheets thus obtained are cut up into flakes, cubes, strips, etc., as required.

Sporting powders are of two kinds, the so-called bulk powders, consisting of loose granules, coated or hardened by means of a solvent, and the so-called condensed powders, gelatinised throughout and made in practically the same way as military flake powders. A powder of an intermediate type, the Walsrode powder, for instance, is made by first gelatinising throughout, and then treating with water or steam. In this way granules are formed and part of the solvent is driven out again, leaving a "bulky" but hard grained powder behind. The bulk powders are supposed to just fill a cartridge used in the old black-powder gun, the condensed powders are made for modern weapons. The usual bulk powders are composed of collodion cotton mixed with potassium or barium nitrate, and generally worked up in an incorporating mill or drum. The mixture is then either sprinkled with water in a rotating drum so as to form grains, or may be pressed and then broken into grains, the solvent being sprinkled over when the grains are already formed.

Smokeless powder gives a strong luminous flame which in military operations will disclose the position of an attacking force. The flame owes its luminosity to incandescent solid particles, and to the sudden combustion, under pressure, of gases which have escaped complete oxidation in the gun, but are still capable of uniting with oxygen. Many attempts to secure absence of flame have been based on the principles applied to the making of safety explosives for coal mines. Sodium bicarbonate and cyanamide, for instance, are added to produce a cooling effect. The following have been the subjects of patents: Vaseline (British Patent, 19,773, 1900); Vaseline and alkali bicarbonate (D.R. Patent, 175,399, 1903); Diethyldiphenylurea (D.R. Patent, 194,874, 1906); Cyanamide (D.R. Patent, 201,215, 1902); Tartrates (British Patent, 15,566, 1905); Olive oil (British Patent, 15,565, 1905); Salts of dibasic organic acids and oils, fats or resins (British Patent, 19,408, 1906); Soaps (D.R. Patent, 195,486, 1907).

The nitro-cellulose used in modern explosives has a tendency to decompose under conditions to which it may frequently be subjected before it is actually used in the gun, and in its gelatinised condition retains small quantities of the solvent which has been used to produce the powder, in spite of the prolonged drying at somewhat elevated temperatures which it has undergone. The removal of residual solvent is important if constant ballistic qualities are to be obtained. Water treatment, followed by drying, has been resorted to in some cases, but wherever heating is necessarily prolonged there is considerable risk of impairing stability.

At the same time it may be pointed out that a heat treatment will sometimes apparently improve a powder, and may actually do so if volatile catalysts alone are in question.

Stability must be carefully tested at regular intervals during storage. Special additions are frequently made during manufacture for the purpose of removing incipient acidity. Aniline, for example, is added to ballistite in Italy, and diphenylamine, which is used in France, acts not only as a so-called "stabiliser," but as an indicator, owing to the change in colour produced by nitrogen oxides. These additions, however, delay deterioration under adverse storage conditions for a short time only.

The composition of a few smokeless powders will be found in the appendix.

The writer of the article desires to acknowledge his indebtedness to the following works :—

"The Manufacture of Explosives," by the late Oscar Guttmann, F.I.C. Whittaker & Co., 1895.
"Twenty Years' Progress in Explosives," by the same author. 1909.
"A Handbook of Modern Explosives," by M. Eissler. Crosby Lockwood & Son, 1897.
"Nitro-Explosives," by P. Gerald Sanford, F.I.C., F.C.S. Crosby Lockwood & Son, 2nd edition, 1906.
"Die Explosivstoffe,"—"Nitroglyzerin und Dynamit" (1908), "Ammonsalpetersprengstoffe" (1909), "Chloratsprengstoffe" (1910); by Dr Richard Escales. Veit & Co., Leipzig.
"New Methods of Testing Explosives," by C. E. Bichel. Cnarles Griffin & Co. Ltd., 1905
Dr Dammer's "Chemische Technologie der Neuzeit." Article on "Explosivstoffe" by Dr Richard Escales. Ferdinand Enke, 1910.
Allen's "Commercial Organic Analysis," new edition. Article by A. Marshall, F.I.C., on "Modern Explosives." 1910.
The Journal of the Chemical Society ; The Journal of the Society of Chemical Industry ; The Chemical Trade Journal ; Zeitschrift für das gesamte Schiess- und Sprengstoffwesen ; Zeitschrift für angewandte Chemie ; Chemiker Zeitung.

Observations on the Testing of Explosives

All materials used in the manufacture of explosives must conform to rigid specifications, which are framed with a view to obtaining a uniform supply free from matters which would affect injuriously both safe and economical working and stability, chemical and physical, of the finished product.

The proximate ingredients of an explosive mixture are subjected to a searching examination before the process of incorporation takes place, and the homogeneous mixture obtained is afterwards tested with regard to (1) proximate composition, (2) chemical stability, (3) capacity for retaining original physical properties under conditions more adverse than those likely to be encountered during storage or shipment, (4) mechanical energy, (5) safety under working conditions, *e.g.*, sensitiveness to shock and friction and liability to ignite mixtures of air and coal-gas, into which sample cartridges are fired.

It is impossible to do more than indicate briefly a few of the more important points connected with the first two tests. For full details the larger works must be consulted.

Proximate Analysis.—Moisture and residual solvent, in the case of nitro-glycerine explosives, are determined by Marshall's modification of L. W. Dupré's method (L. W. Dupré, *Chemiker Zeitung*, 540, 1901, and Marshall, *Journ. Soc. Chem. Ind.*, 154, 1904). P. V. Dupré (*Analyst*, 1906, **31**, 213) estimates water in the volatile matter by passing it through a layer of calcium carbide and measuring the acetylene generated.

Other Constituents.—2 g. of the sample, weighed out into a folded filter paper, are extracted with ether (sp. gr. 0.720) in a Soxhlet apparatus. The extract may contain nitro-glycerine, resin, camphor, sulphur, mineral jelly, or paraffin. In the case of cordite the nitro-glycerine and mineral jelly thus extracted are separated by repeated digestion with aqueous methyl alcohol (80 parts by volume of absolute methyl alcohol to 20 parts by volume of water), which leaves the mineral jelly undissolved. The contents of the filter paper are now extracted with water, and the residue left on evaporating the aqueous solution is examined for ammonium nitrate, potassium chlorate, barium nitrate, and soluble combustible substances. In the insoluble residue aluminium is determined by dissolving it in hydrochloric acid and precipitating it from solution as hydrate. Extraction with acetone will remove nitro-celluloses. The different kinds of carbon may be distinguished in the residue by their behaviour on heating. Wood fibre is recognised by the microscope and by the presence of tannin in an extract, due to bark and nut-galls. Sulphur is transformed into sulphate by oxidation with nitre and potassium chlorate.

Nitro-glycerine, before being used as an ingredient of an explosive mixture, is tested with regard to (1) percentage of nitrogen, (2) alkalinity, (3) moisture, and (4) stability. A direct method of estimating the percentage of nitro-glycerine (usually determined by difference) in an explosive mixture is that of Silberrad, Phillips, and Merriman (*Journ. Soc. Chem. Ind.*, **25**, 628).

Nitro-cellulose is examined for percentage of nitrogen by means of the Lunge nitrometer or by the Schultze-Tiemann process (see Marshall's article, Allen's "Commercial Organic Analysis," 4th edition, 1910, vol. iii.). An important paper on nitrogen estimation is that of Berl and Jurrissen (*Zeits. Schiess-Sprengstoffwesen*, 1910, **4**, 61). Nitro-cellulose is also examined as regards solubility in ether alcohol, amount of unnitrated fibre, matter insoluble in acetone, ash and inorganic matter, degree of alkalinity, presence of sulphuric esters, viscosity of 1 per cent. solutions in ether alcohol or acetone, and stability.

Picric Acid is tested for the presence of resinous and tarry matters, sulphuric acid, hydrochloric acid, oxalic acid, and their salts. General impurities and adulterations, *e.g.*, oxalic acid, nitrates, picrates, boric acid, alum, sugar, etc., remain insoluble on shaking the sample with ether, and may be further examined. The melting point of the pure acid is 122° C. Dinitrophenol may be estimated by Allen's process (*Journ. Soc. Dyers, etc.*, **4**, 84). Mineral or non-combustible matter must not exceed 0.3 per cent. by weight.

Fulminate of Mercury is white or light grey in colour, should dissolve readily and completely, or almost completely, in ammonia or concentrated hydrochloric acid and in potassium cyanide solution. It is examined for metallic mercury, mercurous compounds, and oxalates (see Solonina, *Zeits. Schiess-Sprengstoffwesen*, 1910, **3**, 41, and 1910, **4**, 67).

Cap Composition [$(CNO)_2Hg$; Sb_2S_3; $KClO_3$] is analysed according to Brownsdon's method (*Journ. Soc. Chem. Ind.*, xxiv., April 1905), or that of Messrs F. W. Jones and F. A. Willcox (*Chem. News*, Dec. 11, 1896). The latter dissolve the fulminate by means of acetone saturated with ammonia gas, and afterwards remove the potassium chlorate by extraction with water.

Stability Tests

Most of the numerous tests which have been used or suggested are based upon the detection or measurement of nitrogen oxides evolved when the explosive is heated. They are classified as (1) "Trace Tests," in which the time required to produce a definite colour upon a sensitive test paper suspended over the heated explosive is measured, the temperature being either 160° F. or 180° F.; (2) "Fume Tests," carried out at higher temperatures (100°-135° C.), in which the time elapsing before fumes appear, or before a less sensitive reagent is affected, is measured; and (3) quantitative tests, made at the same temperatures as those attained in "Fume Tests," in which the amount of decomposition products evolved in a given somewhat extended time is determined. No single test has been devised yet which can be relied upon to indicate the probable duration of stability, even under normal conditions. A committee, appointed by the Home Office, has the whole question of the Heat Test under consideration, but has not yet published a report. A bare summary of stability tests follows.

Trace Tests -
1. **Abel Heat Test** (Explosives Act, 1875).—The test paper is impregnated with potassium iodide and starch. (See paper by Robertson and Smart, *Journ. Soc. Chem. Ind.*, Feb. 15, 1910). For detection of mercuric chloride or metallic mercury, the presence of which renders the Heat Test valueless, see method adopted by F. H. and P. V. Dupré, and described in the Annual Report of H.M. Inspectors of Explosives for 1907, p. 17. Hargreaves and Rowe (*Journ. Soc. Chem. Ind.*, 1907, **26**, 813) have worked out a method for the microscopic detection of mercury.
2. Test papers may be impregnated with zinc iodide and starch.
3. **Guttmann's Test** (*Journ. Soc. Chem. Ind.*, 16, 1897).—Diphenylamine sulphate is the indicator used in this case. The test was claimed to be independent of the presence of solvent or of mercuric chloride.
4. **Spica's Test.**—The indicator used is *m*-phenylene diamine hydrochloride (with cane sugar).

Fume Tests -
1. Vieille Test -
2. German 135° Test
See Sy (*Journ. Franklin Inst.*, 1908, 166, pp. 249, 321, 433); also *Journ. Amer. Chem. Soc.*, 1903, **25**, 550.

Explosion Test - The temperature is noted at which 0.1 g. of the sample explodes on being heated at a uniform rate of 5° C. per minute from 100° C.

Waltham Abbey Silvered Vessel Test - See paper by Sir F. L. Nathan, R.A. Also article on "Modern Explosives" by Marshall in Allen's "Commercial Organic Analysis."

Quantitative Tests -
The Will Test.—See Will ("Mittheilungen der Centralstelle," Neu Babelsberg, 1900, 2), and Robertson (*Journ. Soc. Chem. Ind.*, 1901, **20**, 609; *Trans. Chem. Soc.*, 1907, 764, and 1909, 1,241.
Bergmann and Junk Test (*Zeits. Angew. Chem.*, 1904, 982).
Sy's test for nitro-cellulose powders (*Journ. Amer. Chem. Soc.*, 1905, **25**, 550).

Composition of Some Explosives in Common Use

ORDINARY DYNAMITE.

Nitro-glycerine	75 per cent.
Kieselguhr	25 ,,

AMOIS.

Nitrate of ammonia	90 per cent.
Chlor-dinitrobenzene	5 ,,
Wood pulp	5 ,,

AMMONIA NITRATE POWDER.

Nitrate of ammonia	80 per cent.
Chlorate of potash	5 ,,
Nitro-glucose	10 ,,
Coal-tar	5 ,,

CELTITE.

Nitro-glycerine	56.0 to 59.0 parts.
Nitro-cotton	2.0 ,, 3.5 ,,
KNO_3	17.0 ,, 21.0 ,,
Wood meal	8.0 ,, 9.0 ,,
Ammonium oxalate	11.0 ,, 13.0 ,,
Moisture	0.5 ,, 1.5 ,,

ATLAS POWDERS.

Sodium nitrate	2 per cent.
Nitro-glycerine	75 ,,
Wood pulp	21 ,,
Magnesium carbonate	2 ,,

DUALINE.

Nitro-glycerine	50 per cent.
Sawdust	30 ,,
Nitrate of potash	20 ,,

VULCAN POWDER.

Nitro-glycerine	30.0 per cent.
Nitrate of soda	52.5 ,,
Sulphur	7.0 ,,
Charcoal	10.5 ,,

VIGORITE.

Nitro-glycerine	30 per cent.
Nitrate of soda	60 ,,
Charcoal	5 ,,
Sawdust	5 ,,

RENDROCK.

Nitrate of potash	40 per cent.
Nitro-glycerine	40 ,,
Wood pulp	13 ,,
Paraffin or pitch	7 ,,

HERCULES POWDERS.

Nitro-glycerine	75.00 to 40.00 per cent.
Chlorate of potash	1.05 ,, 3.34 ,,
Sugar	1.00 ,, 15.66 ,,
Nitrate of potash	2.10 ,, 31.00 ,,
Carbonate of magnesia	20.85 ,, 10.00 ,,

CARBO-DYNAMITE.

Nitro-glycerine	90 per cent.
Charcoal	10 ,,

GELOXITE (Permitted List).

Nitro-glycerine	64 to 54 parts.
Nitro-cotton	5 ,, 4 ,,
Nitrate of potash	22 ,, 13 ,,
Ammonium oxalate	15 ,, 12 ,,
Red ochre	1 ,, 0 ,,
Wood meal	7 ,, 4 ,,

The wood meal to contain not more than 15 per cent. and not less than 5 per cent. moisture.

GIANT POWDER.

Nitro-glycerine	40 per cent.
Sodium nitrate	40 ,,
Rosin	6 ,,
Sulphur	6 ,,
Kieselguhr	8 ,,

DYNAMITE DE TRAUZL.

Nitro-glycerine	75 parts.
Gun-cotton	25 ,,
Charcoal	2 ,,

RHENISH DYNAMITE.

Solution of N.-G. in naphthalene	75 per cent.
Chalk or barium sulphate	2 ,,
Kieselguhr	23 ,,

AMMONIA DYNAMITE.

Ammonium nitrate	75 parts.
Paraffin	4 ,,
Charcoal	3 ,,
Nitro-glycerine	18 ,,

BLASTING GELATINE.

Nitro-glycerine	93 per cent.
Nitro-cotton	3 to 7 ,,

GELATINE DYNAMITE.

Nitro-glycerine	71 per cent.
Nitro-cotton	6 ,,
Wood pulp	5 ,,
Potassium nitrate	18 ,,

GELIGNITE.

Nitro-glycerine	60 to 61 per cent.
Nitro-cotton	4 ,, 5 ,,
Wood pulp	9 ,, 7 ,,
Potassium nitrate	27 ,,

FORCITE.

Nitro-glycerine	49.0 per cent.
Nitro-cotton	1.0 ,,
Sulphur	1.5 ,,
Tar	10 0 ,,
Sodium nitrate	38.0 ,,
Wood pulp	0.5 ,,

(The N.-G., etc., varies.)

TONITE No. 1.

Gun-cotton	52 to 50 parts.
Barium nitrate	47 ,, 40 ,,

TONITE No. 2.

Contains charcoal also.

TONITE No. 3.

Gun-cotton	18 to 20 per cent.
Ba(NO₃)₂	70 ,, 67 ,,
Dinitro-benzol	11 ,, 31 ,,
Moisture	0.5 ,, 1 ,,

CARBONITE.

Nitro-glycerine	17.76 per cent.
Nitro-benzene	1.70 ,,
Soda	0.42 ,,
KNO₃	34.22 ,,
Ba(NO₃)₂	9.71 ,,
Cellulose	1.55 ,,
Cane Sugar	34.27 ,,
Moisture	0.36 ,,
	99.99

ROBURITE.

Ammonium nitrate	86 per cent.
Chlor-dinitro-benzol	14 ,,

FAVERSHAM POWDER.

Ammonium nitrate	85 per cent.
Dinitro-benzol	10 ,,
Trench's flame-extinguishing compound	5 ,,

FAVIERITE No. 1.

Ammonium nitrate	88 per cent.
Dinitro-naphthalene	12 ,,

FAVIERITE No. 2.

No. 1 powder	90 per cent.
Ammonium chloride	10 ,,

BELLITE.

Ammonium nitrate	5 parts.
Meta-dinitro-benzol	1 ,,

PETROFRACTEUR.

Nitro-benzene	10 per cent.
Chlorate of potash	67 ,,
Nitrate of potash	20 ,,
Sulphide of antimony	3 ,,

SECURITE.

Mixtures of Meta-dinitro-benzol and	26 per cent.
Nitrate of ammonia	74 ,,

RACK-A-ROCK.

Potassium chlorate	79 per cent.
Mononitro-benzene	21 ,,

OXONITE.

Nitric acid (sp. gr. 1.5)	54 parts.
Picric acid	46 ,,

8

EMMENSITE.

Emmens' acid	5 parts.
Ammonium nitrate	5 ,,
Picric acid	6 ,,

BRUGÈRE POWDER.

Ammonium picrate	54 per cent.
Nitrate of potash	46 ,,

DESIGNOLLE'S TORPEDO POWDERS.

Potassium picrate	55 to 50 per cent.
Nitrate of potash	45 ,, 50 ,,

STOWITE.

Nitro-glycerine	58.0 to 61 parts.
Nitro-cotton	4.5 ,, 5 ,,
Potassium nitrate	18.0 ,, 20 ,,
Wood meal	6.0 ,, 7 ,,
Oxalate of ammonia	11.0 ,, 15 ,,

The wood meal shall contain not more than 15 per cent. and not less than 5 per cent. by weight of moisture. The explosive shall be used only when contained in a non-waterproofed wrapper of parchment.—No. 6 detonator.

FAVERSHAM POWDER.

Ammonium nitrate	93 to 87 parts.
Trinitro-toluol	11 ,, 9 ,,
Moisture	1 ,,

KYNITE.

Nitro-glycerine	24.0 to 26.0 parts.
Wood pulp	2.5 ,, 3.5 ,,
Starch	32.5 ,, 3.5 ,,
Barium nitrate	31.5 ,, 34.5 ,,
CaCo₃	0 ,, 0.5 ,,
Moisture	3.0 ,, 6.0 ,,

Must be put up only in waterproof parchment paper, and No. 6 electric detonator used.

REXITE.

Nitro-glycerine	6.5 to 8.5 parts.
Ammonium nitrate	64.0 ,, 68.0 ,,
Sodium nitrate	13.0 ,, 16.0 ,,
Trinitro-toluol	6.5 ,, 8.5 ,,
Wood meal	3.0 ,, 5.0 ,,
Moisture	0.5 ,, 1.4 ,,

Must be contained in waterproof case (stout paper), waterproofed with resin and ceresin.—No. 6 detonator.

WITHNELL POWDER.

Ammonium nitrate	88 to 92 parts.
Trinitro-toluol	4 ,, 6 ,,
Flour (dried at 100° C.)	4 ,, 6 ,,
Moisture	0 ,, 15 ,,

Only to be used when contained in a linen paper cartridge, waterproofed with carnaüba wax, paraffin—No. 7 detonator used.

PHŒNIX POWDER.

Nitro-glycerine	28 to 31 parts.
Nitro-cotton	0 ,, 1 ,,
Potassium nitrate	30 ,, 34 ,,
Wood meal	33 ,, 37 ,,
Moisture	2 ,, 6 ,,

Smokeless Powders

CORDITE.

Nitro-glycerine	-	58 per cent.	+ or −	.75
Nitro-cotton	-	37 ,,	+ or −	.65
Vaseline	-	5 ,,	+ or −	.25

CORDITE M.D.

Nitro-glycerine	-	30 per cent.	+ or −	1
Nitro-cotton	-	65 ,,	+ or −	1
Vaseline	-	5 ,,	+ or −	0.25

Analysis of :—By W. Macnab and A. E. Leighton.

E. C. POWDER.

Nitro-cotton	- - -	79.0 per cent.
Potassium nitrate	- -	4.5 ,,
Barium nitrate	- -	7.5 ,,
Camphor	- - -	4.1 ,,
Wood meal	- - -	3.8 ,,
Volatile matter	- -	1.1 ,,

WALSRODE POWDER.

Nitro-cotton	- - -	98.6 per cent.
Volatile matter	- -	1.4 ,,

KYNOCH'S SMOKELESS.

Nitro-cotton	- - -	52.1 per cent.
Dinitro-toluol	- -	19.5 ,,
Potassium nitrate	- -	1.4 ,,
Barium nitrate	- -	22.2 ,,
Wood meal	- - -	2.7 ,,
Ash	- - -	0.9 ,,
Volatile matter	- -	1.2 ,,

SCHULTZE.

Nitro-ligrin	- - -	62.1 per cent.
Potassium nitrate	- -	1.8 ,,
Barium nitrate	- -	26.1 ,,
Vaseline	- - -	4.9 ,,
Starch	- - -	3.5 ,,
Volatile matter	- -	1.0 ,,

IMPERIAL SCHULTZE.

Nitro-lignin	- - -	80.1 per cent.
Barium nitrate	- -	10 2 ,,
Vaseline	- - -	7.9 ,,
Volatile matter	- -	1.8 ,,

CANNONITE.

Nitro-cotton	- - -	86.4 per cent.
Barium nitrate	- -	5.7 ,,
Vaseline	- - -	2.9 ,,
Lampblack	- - -	1.3 ,,
Potassium ferrocyanide	-	2.4 ,,
Volatile matter	- -	1.3 ,,

AMBERITE.

Nitro-cotton	- - -	71.0 per cent.
Potassium nitrate	- -	1.3 ,,
Barium nitrate	- -	18.6 ,,
Wood meal	- - -	1.4 ,,
Vaseline	- - -	5.8 ,,

SPORTING BALLISTITE.

Nitro-glycerine	- - -	37.6 per cent
Nitro-cotton	- - -	62.3 ,,
Volatile matter	- -	0.1 ,,

CAMBRITE NO. 2.

Nitro-glycerine	-	24-22 parts.
Nitrate of barium	-	4.5-3 ,,
Nitrate of potassium	-	29-26 ,,
Wood meal (dried at 100° C.)	35-32 ,,
Carbonate of calcium	-	1 ,,
Chloride of potassium	-	9-7 ,,
Moisture	- - -	6-3.5 ,,

AMMONITE NO. 1.

Nitrate of ammonium	-	77-73 parts.
Trinitronaphthalene	-	6-4 ,,
Chloride of sodium	-	21.5-19.5 ,,
Moisture	- - -	1-0 ,,

SUPER-RIPPITE.

Nitro-glycerine	-	53-51 parts.
Nitro-cotton	- -	4-2 ,,
Nitrate of potassium	-	15.5-13.5 ,,

Borax (dried at 100° C.)	-	17.5-15.5 parts.
Chloride of potassium	-	9-7 ,,
Moisture	- - -	8-5 ,,

A. 2. MONOBEL.

Nitro-glycerine	- -	11-9 parts.
Nitrate of ammonium	-	61-57 ,,
Wood meal (dried at 100° C.)	10-8 ,,
Carbonate of magnesium	-	1.5-0.5 ,,
Chloride of potassium	-	21.5-18.5 ,,
Moisture	- - -	2-0 ,,

VICTOR POWDER.

Nitro-glycerine	- -	9.5-7.5 parts.
Nitrate of ammonium	-	69.5-66.5 ,,
Wood meal (dried at 100° C.)	9-7 ,,
Chloride of potassium	-	16-14 ,,
Moisture	- - -	2-0 ,,

STATISTICS OF EXPORTS AND IMPORTS OF AMMUNITION, HIGH EXPLOSIVES, ETC., FOR THE UNITED KINGDOM, FOR THE YEARS 1906 AND 1910

EXPORTS.

Material.	1906.		1910.	
	Quantities	Value.	Quantities.	Value.
	Cwt.	£	Cwt.	£
Cordite and other smokeless propellants -	11,894	172,864	12,720	172,399
Gunpowder - - - - -	65,241	176,846	71,788	176,743
High explosives—				
Dynamite and other high explosives -	149,035	898,208	143,931	726,457
Small arms ammunition—				
Military - - - - -	} 84,575	413,530 {	11,454	82,840
Sporting - - - - -			61,117	253,009
Fuzes, tubes, primes, etc. - - - -	1,253	38,866	1,478	57,444
	Thousands.		Thousands.	
Percussion caps - - - - -	281,064	22,712	326,397	24,864
Metal cartridge cases other than small arms ammunition—				
	Cwt.		Cwt.	
Filled - - - - - -	10,122	103,342	13,281	134,803
Empty - - - - - -	789	7,234	2,094	17,449
Shot and shell - - - - -	22,193	110,430	45,959	163,340
Rockets and other combustibles, explosives, and ammunition, of natures not named above - - - - - -	...	301,820	...	407,254
Total for ammunition - - -	...	2,245,852	...	2,216,602

IMPORTS.

Material.	1906.		1910.	
Cordite and other smokeless propellants -	678	10,824	591	8,641
Gunpowder - - - - -	4,744	10,915	4,932	10,934
High explosives—				
Dynamite and other high explosives -	31,774	121,571	9,117	37,748
Small arms ammunition—				
Military - - - - -	} 7,742	50,002 {	24	334
Sporting - - - - -			7,507	44,562
Fuzes, tubes, primes, etc. - - - -	1,943	6,772	2,006	6,173
	Thousands.		Thousands.	
Percussion caps - - - - -	24,571	2,241	16,668	1,962
Metal cartridge cases other than small arms ammunition—				
	Cwt.		Cwt.	
Filled - - - - - -	295	1,538	...	...
Empty - - - - - -	354	1,720	...	...
Shot and shell - - - - -	4,225	21,185	607	2,209
Rockets and other combustibles, explosives, and ammunition, of natures not named above - - - - - -	...	61,036	...	47,658
Total for ammunition - - -	...	287,804	...	160,221

The Imports and Exports of Explosives for the United States are given in the following Statistics

Imports into the United States from All Sources.

Material.	1906.		1909.	
	Quantities, in Lbs.	Value, in Dollars.	Quantities, in Lbs.	Value, in Dollars.
Caps, blasting and percussion (dutiable)	...	14,084	...	8,696
Cartridges (dutiable)	...	124,955	...	68,893
Firecrackers (dutiable)	4,919,122	366,776	5,927,003	343,348
Fireworks (dutiable)	...	21,931	...	52,806
Fulminates and all like articles (dutiable)	...	199,095	...	182,482
Fuse, mining and blasting (dutiable)	...	10,221	...	15,675
All other explosives (dutiable)	...	128,454	...	266,002
				(789,635 in 1910)
Totals	...	865,516	...	937,902

Exports from the United States.

Material.	1906.		1907.		1908.		1909.		1910.	
	Quantities, in Lbs.	Value, in Dollars.	Quantities, in Lbs.	Value, in Dollars.	Quantities, in Lbs.	Value, in Dollars.	Quantities, in Lbs.	Value, in Dollars.	Quantities, in Lbs.	Value, in Dollars.
Cartridges	...	...	...	...	...	1,745,007	...	2,132,394	...	2,521,749
Dynamite					...	793,433	...	615,641	...	1,636,225
Gunpowder	891,379	124,322	472,159	78,620	2,051,356	315,494	690,851	136,82:	1,218,541	249,917
All other explosives (including cartridges and dynamite prior to 1908)	...	3,443,716	...	4,003,782	...	851,583	...	593,857	...	944,772
Totals	...	3,568,038	...	4,082,402	...	3,705,517	...	3,478,714	...	5,352,663

LIST OF PRINCIPAL EXPLOSIVES USED AT MINES AND QUARRIES IN THE UNITED KINGDOM UNDER THE COAL MINES REGULATION ACT, THE METALLIFEROUS MINES ACT, AND THE QUARRIES ACT, IN 1909

PERMITTED EXPLOSIVES.

Name of Explosive.	Quantity Used.	Percentage of Total.	Name of Explosive.	Quantity Used.	Percentage of Total.
				Lbs.	
	Lbs.		Brought forward	8,255,446	97.0
Bobbinite - -	1,118,459	13.2	Celtite - - -	38,639	0.5
Arkite - - -	750,718	8.8	Amvis - - -	34,564	0.4
Monobel - -	739,282	8.7	Geloxite - -	30,527	0.4
Ammonite - -	554,371	6.5	† Steelite - -	29,840	0.3
Samsonite - -	506,991	6.0	Tutol - - -	16,668	0.2
Saxonite - -	504,146	5.9	Fracturite - -	16,382	0.2
Rippite - - -	485,058	5.7	Normanite - -	15,178	0.2
Roburite - -	478,823	5.6	Dominite - -	13,754	0.2
Westfalite - -	444,770	5.2	Minite - - -	9,576	0.1
Bellite - - -	443,181	5.2	Cambrite - -	9,435	0.1
Carbonite - -	431,721	5.1	Aphosite - -	7,600	0.1
Stowite - - -	258,822	3.0	Swalite - -	7,505	0.1
Faversham powder	227,564	2.7	Dragonite - - - 6,050		
* Ammonal - -	214,543	2.5	Amasite - - - 4,000		
Albionite - -	202,769	2.4	Kynite - - - 3,979		
Excellite - -	176,235	2.1	Nobel ammonia powder 1,900		
Rexite - - -	167,164	2.0	Ripping ammonal - 445		
Abbcite - -	120,106	1.4	Odite - - - 396	17,118	0.2
Permonite - -	119,992	1.4	Dahmenite - - 122		
Cornish powder -	99,079	1.2	Electronite - - 100		
Negro powder -	62,566	0.7	Phœnix powder - - 50		
St Helens powder -	60,552	0.7	Britonite - - 40		
Oaklite - - -	45,699	0.5	Withnell powder - 20		
Kolax - - -	42,835	0.5	Haylite - - - 16		
Carried forward -	8,255,446	97.0	Total - -	8,502,232	100.0

* This includes the non-permitted ammonal.
† Includes colliery steelite and non-permitted steelite.

ALL EXPLOSIVES.

Name of Explosive.	Quantity Used.	Percentage of Total.
	Lbs.	
Permitted explosives - - -	8,502,232	28.3
Gunpowder - - - - -	17,595,475	58.5
Gelignite - - - - -	3,085,529	10.2
Blasting gelatine - - - }	616,436	2.0
Gelatine dynamite - - }		
Cheddite - - - -	123,531	0.4
Dynamite - - - - -	117,260	0.4
Various - - - - -	51,424	0.2
Total - -	30,091,887	100.0

ANALYSIS OF EXPLOSIVES[1]

C. G. Storm[2]

The methods described in this chapter cover all of the more common types of explosives employed in the United States for both commercial and military purposes. Those of chief commercial importance are black powder, nitroglycerin dynamites, including "straight" dynamites, ammonia dynamites, gelatin dynamites, and low-freezing dynamites, "Permissible" coal mining explosives and nitrostarch blasting explosives. Military explosives include smokeless powder, guncotton, trinitrotoluene, picric acid, ammonium picrate, "Amatol," tetryl and tetranitroaniline. No sharp distinction can, however, be drawn between commercial and military explosives, as many are utilized for both purposes. Many tons of surplus TNT. have been used recently in commercial work; nitrostarch explosives found important application for military use during the war; mercury fulminate and other detonators and priming compositions are essential in every field of explosives.

The methods described have largely been used by the writer in practical explosives testing and analysis in connection with both Government and private work. Most of those applying to commercial explosives have been approved by the United States Bureau of Mines for use in its Explosives Chemical Laboratory.[3]

BLACK POWDER

The composition of black powder varies to some extent, depending chiefly on the purpose for which the explosive is to be used. Black blasting powder contains sodium nitrate, charcoal and sulphur; black gunpowder is quite similar except that potassium nitrate is generally substituted for the sodium nitrate; black fuse powder is similar to the latter, differing mainly in its granulation. The same general method of analysis is therefore applicable to all types of black powder.

Sampling. From 50 to 100 grams of the original sample is crushed in small portions in a porcelain mortar and completely passed through an 80-mesh sieve, care being taken to avoid undue exposure to the air. The separate powdered portions are promptly bottled and the entire sample is finally well mixed.

Moisture. The standard method of the Bureau of Mines is to desiccate a 2-gram sample on a 3-inch watch glass over sulphuric acid for three days, the loss of weight being moisture. It has been shown, however, that equally accurate results can be obtained by drying at 70° C. in a constant temperature oven to constant weight, for which 2–3 hours is usually sufficient. As much as 5 hours drying at 70° C. will not cause loss of sulphur. Drying at 100° C. gives results which are slightly high, due to loss of sulphur.

[1] Received March, 1920. Published by permission of Chief of Ordnance, U. S. A.
[2] Professor of Chemical Engineering, Ordnance School of Application, Aberdeen Proving Ground, Maryland.
[3] See Bureau of Mines Bulletin No. 51, "The Analysis of Black Powder and Dynamite," W. O. Snelling and C. G. Storm, 1913, and Bulletin No. 96, "The Analysis of Permissible Explosives," C. G. Storm, 1916.

Nitrates. About 10 grams of the finely ground sample in a Gooch crucible provided with an asbestos mat, is extracted with warm water by means of suction, the water being added in 15–20 cc. portions and each portion being allowed to stand in the crucible a short time before suction is applied. About 200 cc. of water is usually sufficient, but the last drops of filtrate should be tested by evaporation to ensure the absence of nitrates. A blue color on the addition of sulphuric acid containing a few crystals of diphenylamine will also indicate the presence of nitrates.

The water extract includes a small amount of water-soluble organic material from the charcoal in addition to the nitrate. It is made up to 250 cc. and an aliquot portion (50 cc.), evaporated to dryness on the steam bath, treated with a little nitric acid, again evaporated, heated to slight fusion and weighed.

If allowance for impurities in the nitrate is desired, a direct determination of nitrate may be made on a separate portion of the water extract by the Devarda method or by means of the nitrometer, but for all practical purposes the evaporation method is sufficient. The usual tests should be made to determine whether sodium nitrate or potassium nitrate is present.

The residue left in the crucible, consisting of sulphur and charcoal, is dried at about 70° C. to constant weight (for 5 hours or over night if more convenient), the loss of weight minus the moisture content being the water-soluble portion. This result serves as a check on the evaporation result.

Sulphur. The residue in the crucible is extracted in a Wiley extractor or other continuous extraction apparatus with carbon disulphide, until evaporation of a small portion of the solvent passing through the crucible shows absence of sulphur. The excess of carbon disulphide is then allowed to evaporate from the crucible in a warm place away from flame, and the residue finally dried to constant weight at 100° C. The loss of weight is considered as sulphur.

Charcoal. The dry residue in the crucible should consist only of charcoal.

Ash. The ash in the charcoal may be determined by ignition over a Bunsen burner until all of the carbon has been burned off, and weighing. This ash also contains, of course, any non-volatile matter that may have been present in the sulphur and nitrate.

Calculation of Results. Since a portion of the charcoal is always dissolved in the water extract, it is customary to express the content of charcoal by subtracting the sum of the following from 100%:

% Moisture (by desiccation or drying at 70° C.).
% Nitrate (by evaporation of water extract with HNO_3).
% Sulphur (by loss on extraction with CS_2).

NITROGLYCERIN DYNAMITES

" Straight " Dynamite

So-called "Straight" nitroglycerin dynamite has been manufactured to only a relatively small extent in this country during the past few years, owing to the high cost of glycerin. It has been largely replaced by the ammonia and low-freezing dynamites, in which a large part of the nitroglycerin is replaced by ammonium nitrate and nitrosubstitution compounds. Furthermore, developments in the manufacture of both ammonium nitrate and nitrocompounds during the war have rendered unlikely any great increase in the manufacture of straight dynamites. They are still largely used, however, where quick-acting blasting explosives of high strength are required, as in work in hard rock. They consist essentially of nitroglycerin absorbed in a "dope" composed of a combustible absorbent, usually wood pulp, and an oxidizing material (sodium nitrate), to which is added a small amount of an antacid (calcium carbonate, zinc oxide, etc.). The analysis is best carried out by successive extractions, usually with ether, water, and dilute hydrochloric acid.

Sampling. The wrappers are removed from a number of the cartridges, and from 3 to 5 cm. of the ends of the exposed roll of explosives rejected. The remainder is thoroughly mixed on a large sheet of paraffined paper or in a large porcelain dish, and an average sample selected and bottled—usually about one half pound. The importance of thorough mixing of the sample must not be overlooked, in view of the fact that there is frequently a decided tendency for the nitroglycerin to segregate due to insufficient or unsuitable absorbent, so that this liquid ingredient may not be uniformly distributed throughout the cartridge. Also if a carefully mixed sample has been allowed to stand for some days, especially in a warm place, segregation may occur in the bottle, so that it is advisable to again mix the sample before analysis.

Qualitative Examination. Although a qualitative analysis of a sample known to be straight nitroglycerin dynamite is usually unnecessary, the exact nature of the sample may be unknown, and a knowledge of the composition of some of the more complex types of dynamite is necessary before a quantitative analysis can be properly conducted.

About 25 grams of the sample is shaken with several successive portions of ether in a large stoppered test tube, the ether being decanted off through a filter paper and the residue finally washed on the filter. The ether solution is allowed to evaporate slowly on a steam bath and the filter paper spread out on a glass plate in an oven so that the residue may dry quickly. The evaporated ether extract may contain nitroglycerin, sulphur (especially in the lower grades of dynamite), rosin, vaseline, or paraffin oil (in ammonia dynamite), nitrotoluenes and other nitrocompounds (in low-freezing dynamites), etc.

Nitroglycerin is readily detected by shaking a drop of the liquid with one or two cc. of strong H_2SO_4 and about 1 cc. of mercury in a test tube, an evolution of brown fumes of nitric oxides being noted if nitroglycerin is present. Sulphur will appear as crystals in the evaporated extract, and may be identified by removing them, washing with acetic acid, and noting the odor of SO_2 on heating in a flame. Rosin, vaseline, oils, etc., appear as a greasy scum on the surface of the nitroglycerin or adhering to the walls of the beaker. These substances, like sulphur, are practically insoluble in acetic acid (70%), and may be separated from the nitroglycerin by means of this solvent. Trinitrotoluene will appear in the nitroglycerin as long yellowish needles, which may be removed, recrystallized from alcohol, and identified by their melting point (approx. 80° C.), or by the red color produced when the alcoholic solution is treated with a little caustic soda solution.

The residue insoluble in ether is replaced in the test tube and treated with water in a similar manner until all water-soluble material has been dissolved. The water solution is tested for sodium, potassium, barium, zinc, etc., and for nitrates, chlorides, etc., using the general methods of qualitative analysis.

The residue is again treated with cold dilute HCl, any effervescence being noted as indicating the presence of a carbonate, and the resulting solution tested for calcium, magnesium, zinc, etc., which may have been present as carbonates or oxides for the purpose of serving as antacids.

The residue insoluble in ether, water, and cold acid may contain wood pulp, starchy cereal products, sawdust, nitrocellulose, ground vegetable ivory (button waste), kieselguhr, ground nut shells, etc. It is most conveniently examined by means of a low-power microscope, whereby its constituents are usually readily determined. Starch is easily detected by heating a portion to boiling with dilute acid, cooling and adding a few drops of iodine solution (in KI), a blue coloration indicating starch.

Moisture. Moisture is best determined by desiccation over sulphuric acid, a sample of about 2 grams being spread evenly over the surface of a 3-inch watch glass and desiccated for 3 days. Continued desiccation causes a gradual loss of nitroglycerin, but the 3-day loss may be safely assumed to closely represent the actual moisture content. The time of the determination may be greatly shortened by the use of a vacuum desiccator, in which case 24 hours desiccation will give a close approximation to the true moisture content.

It must be remembered that in determining moisture in the presence of nitroglycerin, some volatilization of the latter is unavoidable, and that therefore the method followed must be an empirical one. An attempt to desiccate the sample to constant weight will show that there is undoubtedly a continual loss of nitroglycerin. This has been demonstrated[1] by a series of weighings of a sample exposed for a period of 459 days at a constant temperature of 33°–35° C. in an empty desiccator containing no desiccating agent. A gradual loss resulted during the entire period, totaling 17.52% of the original weight of the sample, the original moisture content of which was about 1%.

Extraction with Ether. Ether removes from dynamite not only the nitroglycerin, but, as has already been mentioned, sulphur, resins (present as a component or as a constituent of the wood pulp), oils (usually from cereal products present), etc. Nitrotoluenes, paraffin, vaseline, etc., are not normal constituents of straight dynamite and are considered under the type of explosive in which they are most likely to occur.

Reflux Condenser Method. From 6 to 10 grams of the sample is weighed in either a porcelain Gooch crucible with asbestos mat or a porous alundum filtering crucible of about 25 cc. capacity. The asbestos mat is best prepared as follows: A mixture of 1 liter of water and 5 grams of previously ignited and shredded short fibre asbestos free from hard lumps and very fine material is well shaken and about 10 cc. poured into the crucible. Suction is applied

[1] Storm, C. G., "The Analysis of Permissible Explosives," Bulletin No. 96, Bureau of Mines, pages 21–24, 1916.

and a smooth and perfect mat almost invariably results. The crucibles thus prepared are dried at 100° and are ready for use.

The sample in the extraction crucible is extracted with about 35 cc. of ether (U. S. P.) preferably in a continuous extraction apparatus (Wiley or similar type preferred), for about 45 minutes to 1 hour, water being continuously circulated through the condenser and the extraction tube heated on a water bath, or electric heater, the temperature of which is so regulated that the sample in the crucible will be kept covered with ether without overflowing.

Suction Method. If desired, the ether extraction may be carried out by suction, the Gooch crucible being held in a carbon tube passing through the stopper in a suction flask. About 100 cc. of ether in 6 to 8 portions is passed through the crucible, each portion being allowed to stand in the crucible for one minute before applying gentle suction. No more air than is necessary should be drawn through the sample in order to avoid condensation of moisture in the sample, which might dissolve a portion of the water-soluble salts. This method uses considerably more ether than the reflux condenser method and its chief advantage is that the apparatus required is more simple.

On completion of the extraction the crucible is at once placed in a drying oven, or the excess ether may be removed by suction before drying. If ammonium nitrate is present the drying should be conducted at 70° C. for 18 hours or overnight, but otherwise 5 hours at 100° C. is ample. The loss of weight represents all ether-soluble material plus the moisture in the original sample.

Evaporation of Ether Extract. The ether extract is washed out of the extraction tube or suction flask with a little ether into a tared evaporating dish or small beaker and the ether allowed to evaporate spontaneously in a warm place, or evaporated by means of the "bell jar evaporator."[1] The latter consists of a tubulated bell jar with openings at top and side, placed on a ground glass plate, a slow current of dry compressed air from two drying cylinders containing H_2SO_4 and soda lime respectively, entering the top opening through a glass tube, the lower end of which extends to about one half inch from the surface of the ether solution in the beaker, which is placed on the glass plate. The dry air current striking the surface of the solution with just enough force to cause a slight "dimple," causes rapid evaporation of the ether, and deposition of moisture in the beaker along with the evaporated residue is avoided. The low temperature produced by the rapid evaporation minimizes the loss of nitroglycerin by volatilization. From 5 to 6 hours is usually required for complete evaporation, which should be determined by check weighings. If the bell jar method is not used, the residue, after removal of the ether, must be desiccated over H_2SO_4 for at least 24 hours in order to remove moisture deposited during evaporation.

Nitroglycerin. Nitroglycerin is determined in the dried and weighed ether extract from which all ether has been removed as above described. This determination is best made by means of the du Pont modification of the 5-part Lunge nitrometer (see p. 354, Vol. I). The sample is dissolved in 5–10 cc. of pure sulphuric acid (specific gravity 1.84) and transferred to the generating bulb of the nitrometer, the beaker and cup of the nitrometer being washed with several further additions of acid until a total of 20–25 cc. has been used.

[1] Storm, C. G., "The Analysis of Permissible Explosives," Bulletin, 96, Bureau of Mines, page 35, 1916.

If the quantity of nitroglycerin present is too great, the sample, dissolved in sulphuric acid, is transferred to a burette and an aliquot part run into the nitrometer. The maximum amount of pure nitroglycerin used for the determination should not exceed 0.75 gram. The determination is carried out in the usual manner and the reading of the gas volume in the graduated reading tube divided by .1850 to find the weight of nitroglycerin in the sample used for the determination (pure nitroglycerin contains 18.50% N).

Sulphur, Resins, Oils, etc. It is always preferable to carry out the extraction with ether on duplicate samples, using one sample of the extract for the determination of nitroglycerin as above, and the other for determining sulphur, resins, oils, etc., that may also be contained in the ether extract.

The weighed extract is redissolved in a mixture of ether and alcohol, previously neutralized with standard alkali. The solution thus obtained is titrated with standard alcoholic potash solution using phenolphthalein indicator. 1 cc. of tenth normal alkali is equal to 0.034 grams of rosin (colophony).

A large excess of the alcoholic potash is now added and the mixture heated several hours or overnight on the steam bath to saponify the nitroglycerin. Shake with water and ether in a separatory funnel. The ether solution contains paraffin, vaseline, or mineral oils that may be present, and is evaporated and the residue weighed. The water solution is acidified with HCl, and Br added to oxidize any sulphur. Any separated rosin is filtered off and weighed as a check on the titration, and sulphur determined in the filtrate by precipitation as $BaSO_4$.

Sulphur may also be separated from nitroglycerin by means of acetic acid of approximately 70% strength, the nitroglycerin being quite soluble in acetic acid and the sulphur almost insoluble. The sulphur is filtered from the solution, washed slightly with alcohol to remove the acetic acid solution, dried and weighed.

If a considerable quantity of crystals of sulphur is found in the evaporated ether extract, it is possible that all of the sulphur has not been removed by the ether, and in this case an extraction is made with carbon disulphide, in exactly the same manner as the ether extraction. This extraction is made subsequent to the extraction with water, the sulphur being determined by loss of weight of the residue or by direct weight after evaporation of the carbon disulphide away from free flame.

Extraction with Water and Determination of Nitrates. The dried and weighed residue left in the crucible after extraction with ether, is extracted with water, using a suction flask fitted with a carbon filter tube in which the crucible is held by a short length of thin-walled rubber tubing. Cold water is used for this extraction, as hot water would gelatinize any starch present. A total of at least 200 cc. of water is passed through the sample, in at least 10 portions, each portion being allowed to stand in contact with the residue for a few minutes before being sucked into the flask. An evaporation test of a few drops of the filtrate will determine the completeness of the extraction. When the extraction is complete, the crucible with its insoluble residue is dried for 5 hours, or overnight, at 95°–100° C., and the loss of weight noted as total water-soluble material. This includes nitrates and other soluble salts that may be present, together with water extract from the wood pulp, flour or other absorbent. This soluble organic material may amount to as much as 2% of the total sample, when cereal products are present. Calcium, magnesium, or zinc may also be present in solution, resulting from the action of acid decomposition products of the nitroglycerin on the carbonate or other antacid present. In routine analyses of ordinary dynamite, the loss of weight on extraction with water is usually considered as the alkaline nitrate (sodium or potassium), but where more exact results are desired an aliquot portion of the extract is evaporated to dryness with a little nitric acid to oxidize organic materials, and the residue weighed as alkaline nitrate. This weight may be corrected for inorganic impurities—chlorides, sulphates, iron, aluminum, calcium, etc.—determined separately by the usual methods.

Nitrates may be determined by means of the nitrometer, using an aliquot portion of water extract estimated to contain .6 to .8 gram of $NaNO_3$ or .8 to 1.0 gram of KNO_3. This is evaporated on the steam bath almost to dryness and transferred with as little water as possible, to the cup of the nitrometer. This solution is drawn into the generator and 30 to 40 cc. of 95–96% H_2SO_4 added slowly so as to avoid generating sufficient heat to crack the glass. The generator is then shaken for a total time of 8–10 minutes in order to be certain that the generation of gas is complete with the diluted acid. The gas is measured and the % of nitrate calculated as in the case of nitroglycerin.

Extraction with Acid. When starch is not present in the residue, a simple extraction of the residue insoluble in water is made with cold dilute HCl (1 : 10), 100 cc. being drawn through the sample in the crucible in small successive portions as described under "Extraction with Water." Several portions of water are then drawn through to wash out the acid, and the residue in the crucible dried for 5 hours at 95° to 100° C. The loss of weight is usually reported as antacid, but the base dissolved may be determined by the usual quantitative methods if desired. The acid-soluble materials generally present are calcium or magnesium carbonate or zinc oxide.

Determination of Starch. If starch is present in the residue insoluble in water, it is removed together with the antacid by boiling with dilute acid. The residue is moistened with water, scraped or washed out of the crucible into a 500 cc. beaker, the volume brought to about 250 cc. by the addition of water and 3 cc. of strong HCl, and the mixture boiled until a drop of the solution fails to give a blue color when treated on a spot plate with a drop of a solution of iodine in KI. This indicates that the starch has been completely hydrolyzed to dextrin. The mixture is then filtered through a fresh crucible, washed with water, dried and weighed, correction being made for the weight of the asbestos mat of the original crucible.

The antacid dissolved in the acid filtrate is determined as already described. The loss of weight by the boiling treatment, minus the antacid found, represents starch and other dissolved organic materials removed from cereal products or wood pulp. The insoluble residue includes the wood pulp and the crude fibre of the cereal products.

Because of the impracticability of exact separations it is customary to report all of the soluble organic material included in both water and acid extractions as "starch" or "starchy material," and the insoluble organic residue as "wood pulp and crude fibre," or the sum of these organic materials is often reported as "carbonaceous combustible material."

Insoluble Residue and Ash. The insoluble residue may contain wood pulp or sawdust, the crude fibre from various cereal products such as corn meal, wheat flour, middlings, bran, etc., ground nut shells, vegetable ivory meal, and more rarely inorganic material such as infusorial earth (kieselguhr), clay, etc. These can usually be identified by microscopic examination (see Bureau of Mines Bulletin 96, Page 74), and a determination of the ash will show whether inorganic materials are present. A high ash content may also indicate incomplete water or acid extractions.

Ammonia Dynamite

So-called ammonia dynamite is essentially "straight" dynamite in which a large part of the nitroglycerin is replaced by ammonium nitrate. The ammonium nitrate is frequently protected from moisture by a coating of vaseline or paraffin and is usually neutralized with zinc oxide. This type of dynamite generally contains less wood pulp than the corresponding grades of "straight" dynamite, and sulphur and cereal products, such as low grade flour, are usually present.

The determination of moisture and the various extractions are carried out as described for "straight" dynamite. An extraction with carbon disulphide is usually necessary to effect complete removal of the sulphur; this properly follows the extraction with water. · The analysis of the ether extract may be conducted as already described. In drying the residue left in the crucible after extraction with ether, it is important that a temperature of approximately 70° C. be used, because in the presence of ZnO, the loss of ammonium nitrate is considerable at 100° C. Pure ammonium nitrate is not appreciably affected by even 24 hours heating at 100° C., but the presence of the ZnO causes decomposition at this temperature.

The water extract contains sodium nitrate and ammonium nitrate together with practically all of the zinc oxide present, the latter ingredient being dissolved with the ammonium nitrate, and a small amount of soluble organic material from the flour or other absorbent. It is analyzed as follows: An aliquot portion is evaporated to dryness in a platinum or silica dish on a steam bath, the ammonium nitrate volatilized by careful heating over a burner, a little nitric acid added to re-oxidize any nitrate that may have been reduced to nitrite, and the residue again dried on the steam bath. The zinc oxide is now in the form of zinc nitrate and may be separated from the sodium nitrate by either of the following methods:

1. The residue is dried at 110°–120° C. and weighed as $NaNO_3$ and $Zn(NO_3)_2$. It is then dissolved in water, the zinc precipitated with Na_2CO_3, filtered, ignited and weighed as ZnO, and the $NaNO_3$ taken by difference; the total $NaNO_3$ plus $Zn(NO_3)_2$ minus $(ZnO \times 2.33) = NaNO_3$.

2. The residue is gently heated over a burner until evolution of oxides of nitrogen from decomposition of the $Zn(NO_3)_2$ has ceased, and the remaining residue weighed as $NaNO_3$ and ZnO. It is then treated with water, the insoluble ZnO filtered on a Gooch crucible, ignited and weighed, the $NaNO_3$ being taken by difference.

Ammonium nitrate is determined in a separate portion of the water extract by the usual method of distillation and titration.

The sum of the amounts of NH_4NO_3, $NaNO_3$, and ZnO found will be somewhat less than the total water extract owing to the presence of water-soluble organic material from the carbonaceous absorbents.

Gelatin Dynamite

This is a form of nitroglycerin explosive in which the nitroglycerin, instead of being absorbed in porous materials such as wood pulp, is combined with nitrocellulose in the form of a gelatinous plastic mass. As little as 3.5% of suitable grade of nitrocellulose containing about 12% nitrogen will, when heated with nitroglycerin, at about 60° C., form a jelly-like non-fluid mass when cooled to ordinary temperature. "Blasting gelatin," used to a considerable extent where great strength is required, is a stiff colloid composed of 90 to 93% nitroglycerin and 10 to 7% nitrocellulose.

All blasting explosives containing such colloids of nitroglycerin and nitrocellulose combined with an active "dope" or base, consisting of a nitrate and combustible material, are termed gelatin dynamites. This type of explosive is also known in some countries as "Gelignite."

Sampling. Owing to its pasty consistency the sample of gelatin dynamite must be prepared by cutting portions of a number of cartridges into thin bits with an aluminium or platinum spatula. The use of a steel spatula or knife for this purpose is not to be recommended for reasons of safety. An ample quantity of sample thus prepared is well mixed and bottled. Owing to its tendency to again form a solid mass upon standing, it should be analyzed as soon as possible after being prepared.

Analysis. The principal ingredients that may be found in the different types of gelatin dynamite are nitroglycerin; nitrocellulose; sulphur; rosin; sodium, potassium or ammonium nitrate; calcium or magnesium carbonate; wood pulp, cereal products and similar carbonaceous combustible materials. Low-freezing gelatins may also contain nitrotoluenes or other nitrosubstitution compounds.

Moisture is determined as described for "straight" dynamite, and the extraction with ether made in the usual manner except that ether free from alcohol (distilled over sodium) is used in order to prevent partial solution of the nitrocellulose. The latter is readily soluble in a mixture of ether and alcohol, and as ordinary U. S. P. ether contains about 4% of alcohol, there is a possibility that an appreciable part of the 0.5% to 2.0% of nitrocellulose present in the sample will be dissolved unless pure ether is used. The ether extract is evaporated and analyzed as already described and the water extraction made in the usual manner. If more than 1 or 2% of sulphur was present it will not have been completely removed by the ether, unless the extraction was continued for a sufficiently long time. In this event, it is necessary to make an additional extraction with carbon disulphide in the Wiley apparatus subsequent to extraction with water.

Nitrocellulose. After the extractions with ether, water, and CS_2 (if necessary) have been made, the nitrocellulose is determined, preferably by extraction with acetone, which is a better solvent for the purpose than a mixture of ether and alcohol. It is advisable to separate the dry residue from the crucible, leaving the asbestos mat intact if possible. The residue is transferred to a small beaker, covered with acetone and allowed to stand at least 3 or 4 hours with occasional stirring. It is then filtered through the original crucible, washed with acetone, dried and weighed, the loss of weight being regarded as nitrocellulose. To correct for small amounts of extract from the wood pulp or other carbonaceous material, the acetone solution may be evaporated to about 20–25 cc., and diluted gradually with a large volume (about 100 cc.) of hot water, which volatilizes the acetone, precipitating the nitrocellulose as a white flocculent mass, which is filtered, dried, and weighed.

The remainder of the analysis is conducted as for straight dynamite.

It will be found that the results of analysis of a gelatin dynamite do not agree with its trade markings. For example, the usual "40% strength" gelatin dynamite actually contains from 30 to 33% of nitroglycerin and about 1% of nitrocellulose. Weight for weight this explosive is considerably weaker than 40% straight dynamite, which contains 40% of nitroglycerin.

Low=Freezing Dynamite

Low-freezing dynamites vary from the dynamite types already discussed by containing an ingredient which reduces the freezing point of the nitroglycerin. This ingredient replaces a portion of the nitroglycerin which would be used in an equal grade of ordinary straight dynamite, ammonia dynamite, or gelatin dynamite. While straight nitroglycerin dynamite may freeze at temperatures as high as 8° C. (46° F.), some of the low-freezing dynamites freeze only at temperatures considerably below 0° C. Many of this type, however, cannot be relied upon to resist freezing at temperatures below the freezing point of water.

The additions made to nitroglycerin for this purpose include the nitrotoluenes, nitroxylenes, nitrohydrins, nitrosugar, and nitropolyglycerin (tetranitrodiglycern). Any of these substances present will be found in the ether extract together with, and in most cases dissolved in, the nitroglycerin.

Moisture. The determination of moisture is carried out as already described for "straight" nitroglycerin dynamite (p. 1375). Attention has been called to the fact that certain nitrosubstitution compounds, notably the mono- and dinitrotoluenes, are more or less volatile and would therefore be partly lost if the moisture is determined in a vacuum desiccator. The safest procedure is therefore to determine the moisture by desiccation for 3 days without vacuum. The difference between the total loss on extraction with ether and the direct weight of the ether extract, after evaporation of the ether in a bell-jar evaporator (p. 1376), should be equal to the moisture content of the sample. This figure will therefore serve as a check on the result obtained by desiccation.

Nitrotoluenes. Trinitrotoluene is not readily soluble in nitroglycerin and separates as crystals on evaporation of the ether from the ether extract, enabling it to be qualitatively separated and identified. It may be determined by difference, the nitroglycerin being determined by means of the nitrometer. Any dinitrotoluene present may also be determined in this manner together with the trinitrotoluene, but if mononitrotoluene is also present, the determination of the nitrogen of the nitroglycerin will be slightly in error by about 0.5530 gram of nitroglycerin for every gram of mononitrotoluene present.[1]

Mononitrotoluene is, however, seldom present except as an impurity in the so-called liquid di- and trinitrotoluenes used in low freezing dynamites, so that the determination of the nitroglycerin is usually fairly accurate and the nitrotoluenes may be calculated by difference.

[1] Storm, C. G., "The Effect of Nitrotoluenes on the Determination of Nitroglycerin by Means of the Nitrometer," Proc. 8th Int. Cong. Appl. Chem., Vol. 4, 1912, p. 117; also Bu. of Mines Bull. 41, p. 62, 1913.

The total nitrogen of the combined nitroglycerin and nitrosubstitution compound may also be determined, the nitrogen of the nitroglycerin deducted and the amount of nitrosubstitution compound calculated from the resulting difference, if the identity of the nitrosubstitution compound has been established. A suitable modification of the Kjeldahl method which has been found applicable to difficultly decomposable nitrocompounds is as follows:[1] This method is, of course, applicable to mixtures containing nitroglycerin.

Modified Kjeldahl Method for Nitrogen. About 0.5000 g. of the nitrocompound is weighed into a 500 cc. Kjeldahl flask, 30 cc. of 96% H_2SO_4 and 2 g. salicylic acid added and the sample dissolved by heating on a steam bath if necessary. Cool; add 2 g. zinc dust in small portions, with cooling and rotating the flask. Continue the shaking at 15 minute intervals for 2 hours and let stand overnight. Then heat over a small flame till fuming has ceased (about 2 hours), cool slightly and add 1 g. HgO. and boil $1-1\frac{1}{2}$ hours longer. Cool and add 7.5 g. K_2SO_4 and 10 cc. H_2SO_4 and boil $1\frac{1}{2}$ to 2 hours more. If the solution is not clear and almost colorless, add 1 g. more K_2SO_4 and boil longer. Cool and add 250 cc. H_2O to dissolve the cake formed, then add 25 cc. K_2S solution (80 g. per liter H_2O), 1 g. granulated Zn, and 85–90 cc. NaOH solution (750 g. per liter H_2O), and distill as usual in the Kjeldahl determination, collecting the NH_3 in standard H_2SO_4 solution. A blank determination without sample is advisable.

Separation of Nitrocompounds from Nitroglycerin. Hyde has devised a satisfactory method for actual separation of nitrosubstitution compounds from nitroglycerin, depending on the differences in solubility of these ingredients in carbon bisulphide and dilute acetic acid.[2] Nitroglycerin is only slightly soluble in CS_2, but readily soluble in dilute acetic acid, while most nitrocompounds are much more soluble in CS_2 and much less soluble in dilute acetic acid than nitroglycerin. CS_2 and acetic acid are only slightly miscible. Hence nitroglycerin and a nitrocompound may be partly separated by shaking the mixture with CS_2 and dilute acetic acid, allowing the two solvents to separate into two layers and drawing off one of the layers. The CS_2 layer will contain most of the nitrocompound and the acetic acid layer most of the nitroglycerin.

Hyde's method involves a continuous fractional extraction in a rather complicated apparatus consisting of 13 long narrow extraction tubes, connected with each other and with a condenser, reservoir and distilling flask so as to form a closed circulating system, the CS_2 continually passing in a train of fine drops through acetic acid in the series of extraction tubes, carrying with it the nitrocompound, the nitroglycerin tending to remain dissolved in the acetic acid. Practically a complete separation is finally obtained, the nitrocompound dissolved in the CS_2 collecting in the distilling flask at the end of the extraction train and the nitroglycerin remaining in solution in the acetic acid in the tubes. The CS_2 is evaporated and the nitrocompound weighed. Reference should be made to the original article by Hyde for details as to construction and operation of the apparatus.

[1] Cope, W. C., "Kjeldahl Modification for Determination of Nitrogen in Nitrosubstitution Compounds," J. Ind. and Eng. Chem., Vol. 8, p. 592, 1916.

[2] Hyde, A. L., "The Quantitative Separation of Nitrosubstitution Compounds from Nitroglycerin," J. Am. Chem., Soc., Vol. 35, p. 1173, 1913. (See also Bu. Mines Bulletin, 96, pp. 47–50, 1916.)

Nitrosugars. The nitrates of sugar, improperly called nitrosugar, are used to a considerable extent for lowering the freezing point of nitroglycerin. This substance is soluble in nitroglycerin, being prepared with the latter by nitrating a solution of cane-sugar in glycerin, and no method is known for its separation from nitroglycerin. Hoffman and Hawse[1] have reported on an optical method for the determination of nitrated sugar in nitroglycerin mixtures, based on the use of the polariscope. As an example of the application of the method, 10.65 g. of a nitrated mixture of glycerin and sugar was dissolved in 100 cc. alcohol and its angle of rotation found to be $a = 3.07°$. The specific rotatory power of sucrose octanitrate having been determined as $\alpha = 56.66$, the formula: C (concentration) $= a/2\alpha$ gives a result of 25.44% sucrose octanitrate in the sample.

The result of the optical method may be roughly checked by a determination of the total nitrogen of the combined nitroglycerin and nitrosugar, assuming the nitrogen content of the nitrosugar to be 15% (theoretical for sucrose octanitrate 15.95%), and that of nitroglycerin 18.50%.

Nitrochlorhydrins. Dinitromonochlorhydrin has been known for years as a partial substitute for nitroglycerin in explosives. It is a solvent for nitrocellulose in smokeless powders and has an appreciable effect in lowering the freezing point of nitroglycerin. During recent years it has come into use in this country as a substitute for nitrotoluenes in low freezing dynamites.

A mixture of dinitrochlorhydrin and nitroglycerin will have a lower nitrogen content than pure nitroglycerin, the dinitrochlorhydrin containing only 14.0% N, as compared with 18.50% N in nitroglycerin. The dinitrochlorhydrin may be readily identified and determined quantitatively by treating the mixture containing this substance and nitroglycerin with an excess of alcoholic solution of KOH, heating on the steam bath until saponification is complete, and determining the chlorine in the solution as chloride.

It must be noted that dinitrochlorhydrin is somewhat more volatile than nitroglycerin and therefore in evaporating the ether from the ether extract it is advisable to make use of the bell-jar evaporator (p. 1376) so as to minimize its loss during evaporation.

Nitropolyglycerin. Nitrated polymerized glycerin—usually a mixture of tetranitrodiglycerin and trinitroglycerin—is sometimes found in low-freezing explosives. This mixture will show a lower N-content than nitroglycerin, since pure tetranitrodiglycerin contains only 16.19 % N. The presence of the latter substance is indicated by low solubility in dilute acetic acid (60 volumes glacial acetic acid to 40 volumes water). One gram of nitroglycerin dissolves in about 10.5 cc. of this acid, while 1 gram of a mixture containing 82.25% tetranitrodiglycerin required 120 cc. of the acetic acid to completely dissolve it. In dissolving such a mixture, it will be found that a part of the mixture dissolves more readily than the remainder. If the more difficultly soluble portion is separated, dried in a desiccator and its nitrogen content determined in the nitrometer, it will be found to contain a much lower % N than the original mixture, approximating the figure for tetranitrodiglycerin, 16.19% (an actual trial gave 16.24% N).

If the presence of tetranitrodiglycerin is established by the above procedure and no other substances except nitroglycerin are present, the proportions of these two ingredients in the ether extract may be readily calculated from the N-content as found by the nitrometer.

[1] Hoffman, E. J. and Hawse, V. P., "The Nitration of Sucrose Octanitrate," J. Am. Chem. Soc., Vol. 41, pp. 235–247, 1919.

"PERMISSIBLE" EXPLOSIVES

"Permissible" explosives are coal mining explosives which have passed the prescribed tests of the Bureau of Mines and are recommended by the Bureau for use in gassy and dusty mines. Their important characteristic is a relatively low flame temperature, which is brought about by modifying the composition of the usual types of dynamites and other blasting explosives. The general methods of reducing the flame temperature of explosives[1] are summarized as follows:

(*a*) Addition of an excess of carbon,—forming less CO_2 and more CO in the gases of explosion.

(*b*) Addition of free water or of solids with water of crystallization.

(*c*) Addition of inert materials.

(*d*) Addition of volatile salts.

The analysis of explosives of this class is therefore generally more complicated than that of the ordinary types of blasting explosives because of the greater variety of ingredients used in manufacture. A partial list of substances which have been found in low-flame explosives manufactured in this country is shown below, arranged according to their solubility in the general scheme of analysis:

Soluble in Ether
Nitroglycerin
Nitropolyglycerin
Nitrotoluenes
Nitrosugars
Nitrochlorhydrins
Paraffin
Resins
Sulphur
Vaseline
Oils

Soluble in Water
Ammonium nitrate
    "   chloride
    "   sulphate
    "   oxalate
    "   perchlorate
Alum (cryst.)
Aluminum sulphate (cryst.)
Barium nitrate
Calcium sulphate (cryst.)
Gums
Magnesium sulphate (cryst.)
Potassium chlorate
    "   nitrate
    "   perchlorate
Sodium nitrate
    "   chloride
    "   bicarbonate
    "   carbonate
Sugar
Zinc oxide

Soluble in Acids
Aluminum
Calcium carbonate
    "   silicide
Ferric oxide
Magnesium carbonate
Zinc
Zinc oxide

Insoluble

Charcoal
Clay
Coal
Corn meal
Corncob meal
Kieselguhr
Nitrocellulose
Nitrostarch
Nitrated wood

Peanut shell meal
Powdered slate
Rice hulls
Sawdust
Turmeric
Vegetable ivory meal
Wheat flour
Wood pulp

[1] The thermochemical considerations involved are discussed in Bureau of Mines Bulletin No. 15, "Investigations of Explosives used in Coal Mines," 1912, and the details of analysis in Bureau of Mines Bulletin No. 96, "The Analysis of Permissible Explosives," C. G. Storm, 1916.

Qualitative Analysis. The qualitative examination of a "permissible" explosive is conducted in the same manner as has been described for dynamite (see page 1374), and, in view of the greater variety of constituents that may be present, is quite essential before a suitable scheme for quantitative separation can be chosen.

Tests for some of the more unusual substances not generally found in the ordinary types of blasting explosives, and not already discussed under "Low-freezing Explosives," are made as follows:

Test for Sugar. The presence of water-soluble organic substances is indicated by an appreciable charring of the residue obtained by evaporating a portion of the water extract to dryness and then heating gradually over a burner. A slight charring may result from water-soluble portions of cereal products, wood-pulp, etc., and may be disregarded. Sugar is identified by acidifying some of the water solution with a little dilute HCl, heating to boiling, neutralizing with KOH and then boiling with Fehling's solution. A precipitation of cuprous oxide indicates the presence of sugar.

Test for Gum Arabic. Gum arabic is precipitated by the addition of a solution of basic lead acetate to the water extract, a white, flocculent precipitate of indefinite composition resulting (see Determination of Gum Arabic, p. 1388).

Test for Nitrostarch. Nitrostarch is best identified by microscopic examination of the residue insoluble in water. It is easily distinguished from unnitrated starch by means of a solution of iodine in KI, which colors the starch granules dark blue but does not affect the granules of nitrostarch.

Test for Chlorides, Chlorates, and Perchlorates. These three substances present in a solution may be identified as follows: Acidify slightly with nitric acid, add excess of AgNO₃, heat to boiling, shake well, and filter off the silver chloride. To the filtrate add a few cc. of 40% solution of formaldehyde (formalin), and boil to reduce chlorates to chlorides. This reduction is best carried out by heating on the steam bath for about an hour. Any chloride thus formed is then separated by further precipitation with AgNO₃ and removed by filtration. The filtrate is then evaporated to dryness, the residue transferred to a crucible and fused with dry Na₂CO₃. The fused mass is treated with dilute HNO₃, when the presence of perchlorate will be indicated by an insoluble precipitate of AgCl.

Mechanical Separation of Solid Ingredients. It is frequently of advantage, especially in connection with the interpretation of the results of analysis of an explosive mixture containing a number of water-soluble salts, to determine the identity of one or more of the components of the mixture by means of screening or by a method of separation depending on variation in specific gravity of the components. Such methods are facilitated by the fact that the ingredients of blasting explosives are frequently not finely pulverized in the course of manufacture.

(a) *By Screening.* 25 to 50 grams of the sample is washed several times with ether to remove nitroglycerin and ingredients of an oily nature, the solid residue dried to remove adhering ether and then sifted through a set of sieves. An examination of the portions held by the 10- and 20-mesh screens will usually show the presence of coarse crystals which are large enough to be sorted out with the aid of forceps, submitted to qualitative tests and identified with certainty. A single crystal may sometimes be identified by dissolving it in a drop of water on a microscope slide, allowing the water to evaporate and examining the resulting crystals under the microscope. The writer has frequently identified three or four ingredients of an explosive in this manner.

(b) *By Specific Gravity Separations.* This method, applied to the analysis of explosives by Storm and Hyde,[1] depends on the separation of solids from a mixture by means of inert liquids of different specific gravities. A series of mixtures of chloroform (sp.gr. 1.49) and bromoform (sp.gr. 2.83) is prepared covering as wide a range of specific gravity as may seem desirable. Portions of the dried sample previously extracted with ether as in (a) are added to such liquid mixtures and the heavier salts, which settle to the bottom, separated from the lighter ones. For example a mixture of ammonium nitrate (sp.gr. 1.74) and sodium chloride (sp.gr. 2.17) is readily separated into its components in a liquid with a specific gravity of (*e.g.*), 1.90, so that the components can be tested separately and the analyst assured that the mixture is not composed of sodium nitrate and ammonium chloride,—which could not be ascertained by ordinary quantitative analysis. (For example, a mixture composed of 16.61% Na, 44.76% NO₃, 13.00% NH₄ and 25.63% Cl may contain either 61.37% NaNO₃ and 38.63% NH₄Cl, or 57.76% NH₄NO₃ and 42.24% NaCl, or varying proportions of all four ingredients.) The chloroform-bromoform mixtures are recovered by filtering and used repeatedly.

The specific gravities of some of the more common salts that may be found are as follows:

Ammonium alum (cryst.)............................1.62
"   chloride...............................1.52
"   nitrate................................1.74
"   perchlorate............................1.87
"   sulphate...............................1.77
Barium nitrate.....................................3.23
Calcium carbonate (ppt'd)..........................2.72
"   sulphate (anhydrous)......................2.97
"   sulphate+2H₂O.............................2.32
Magnesium carbonate................................3.04
"   sulphate+7H₂O...........................1.68
Manganese dioxide..................................5.03
Potassium alum (cryst.)............................1.75
"   chlorate...............................2.33
"   chloride...............................1.99
"   nitrate................................2.09
"   perchlorate............................2.52
"   sulphate...............................2.66
Sodium chloride....................................2.17
"   nitrate..................................2.26
"   sulphate (anhydrous).....................2.66
"   sulphate+10H₂O...........................1.46

Moisture. The determination of moisture in all types of "permissible" explosives is carried out by the method described for nitroglycerin dynamites (page 1375). The influence of the slight volatility of nitroglycerin and of certain nitrosubstitution compounds on the results of this determination has been discussed (pp. 1375, 1381). A more serious factor in the case of many "permissible" explosives is the presence of salts containing water of crystallization. Most salts of this type (*e.g.*, MgSO₄.7H₂O) undergo a gradual loss of a large part of their combined water on desiccation over either H₂SO₄ or CaCl₂, thus rendering it impossible to differentiate between hygroscopic moisture and

[1] Storm, C. G., and Hyde, A. L., "Specific Gravity Separation Applied to the Analysis of Mining Explosives," Tech. Paper No. 78, Bureau of Mines, 1914.

combined water. Attempts to remove the total water content by heating at a temperature high enough to drive off all of the water of crystallization are useless on account of the increased volatilization of nitroglycerin, ammonium nitrate, etc., at such temperatures.

In such cases it is necessary to determine all other constituents by direct methods and estimate moisture by difference, the salt to which the water of crystallization belongs being calculated as containing its full quota of water; or the crystallized salt may be calculated as anhydrous and the difference from 100% reported as "water of crystallization plus moisture."

Extraction with Ether. The extraction with ether, the evaporation of the ether, and the analysis of the ether-soluble portion are conducted as already discussed for nitroglycerin dynamites (pp. 1375, 1376).

In drying the crucibles containing the residue insoluble in ether, a temperature of 100° C. may be used except when the residue contains ammonium nitrate or organic nitrates such as nitrocellulose, nitrostarch, or nitrated wood. When any of these substances are present, the residue should be dried to constant weight at 70° C. Except when salts containing water of crystallization are present, the amount of ether-soluble material found is calculated by deducting the moisture determined by desiccation, from the difference between the weight of original sample and the weight of the dried residue insoluble in ether. The procedure followed when water of crystallization is present is noted in the preceding paragraph.

Extraction with Water. Water-soluble salts are extracted from the weighed residue insoluble in ether as already described, the residue left in the crucibles dried to constant weight at 95 to 100° C., cooled and weighed. The water soluble salts in the solution are determined by the usual methods of inorganic analysis.

Nitrates. In determining nitrates by the nitrometer method (see p. 354) it must be remembered that the presence of a considerable quantity of chlorides may interfere with the accuracy of the results. Many of the "permissible" explosives contain sodium chloride in amounts varying from 1% to 10 or 15%. M. T. Sanders[1] has shown that if the sodium chloride is present in an amount exceeding 15–17% of the sodium nitrate, the result is not accurate within 0.1%. Smaller amounts of sodium chloride do not interfere, except to increase the amount of sludge formed in the nitrometer.

Nitrates may also be determined by the "nitron" method of Busch.[2]

Chlorates. Chlorates may be determined by any of the methods described on page 152 (reduction with SO_2, $FeSO_4$, or Zn) or by the formaldehyde method.[3] In the latter method a portion of the solution, containing about 0.5 g. of chlorate is diluted to 150 cc., 5–10 c. of 40% formaldehyde solution, 2 cc. dilute HNO_3 (1 : 3), and 50 cc. of approx. tenth normal silver nitrate added, the solution covered and heated on the steam bath for about 4 hours, when the precipitate of AgCl is filtered off, washed, dried and weighed. This method is accurate to .05 to .10%.

[1] Sanders, M. T., "The Effect of Chlorides on the Nitrometer Determination of Nitrates," J. Ind. & Eng. Chem., 12, p. 169–170, 1919.

[2] See page 345, also, for further details, "The Analysis of Permissible Explosives," Bureau of Mines, Bulletin 96, pages 60–2.

[3] Storm, C. G., "The Analysis of Permissible Explosives," Bulletin 96, Bureau of Mines, pp. 63–4, 1916.

Perchlorates. The determination of perchlorates by reduction to chlorides on ignition with NH_4Cl in the presence of platinum is described on page 1128. Perchlorates may also be determined by means of precipitation with "nitron" in exactly the same manner as for nitrates. The weight of nitron perchlorate $(C_{20}H_{16}N_4HClO_4)$ found, multiplied by 117.5 (mol. wt. of NH_4ClO_4) and divided by 412.5 (mol. wt. of nitronperchlorate) gives the weight of perchlorate found, expressed as NH_4ClO_4.

Gum Arabic. This substance, sometimes used as a binder in dry explosive mixtures—especially chlorate or perchlorate powders—is determined by precipitation with basic lead acetate solution, prepared by adding 150 g. of normal lead acetate and 50 g. lead oxide (PbO) to 500 cc. distilled water, heating almost to boiling, and filtering. This reagent is added to the solution containing the gum arabic until no further precipitation occurs; the mixture is allowed to stand for several hours, then filtered, washed with absolute alcohol, dried at 100° and weighed. The weight of precipitate multiplied by the factor 0.4971 (determined experimentally) gives the weight of gum arabic found. Chlorides or sulphates, if present, interfere with the determination and must be first removed.

Sugar. Sugar may be present as an ingredient in some "permissible" explosives, and is always found in small amounts in the water extract if cereal products such as corn meal or wheat middlings are present. A portion of the water extract is acidified with HCl (1 cc. conc. HCl to 100 cc. solution), heated just to boiling, cooled, nearly neutralized with Na_2CO_3, an excess of Fehling's solution added and the mixture heated until reduction is complete. The Cu_2O is filtered from the blue liquid, dried, ignited to constant weight, and weighed as CuO. This weight $\times 0.4308$ equals weight of cane sugar. The result is corrected for the result of a blank determination using distilled water instead of the water extract. By the use of this method after first extracting with ether, then with water, corn meal was found to contain 2.65% and wheat middlings 6.25–7.00% of sugar. Thus an explosive containing 25% wheat middlings would show as much as 1.75% of sugar in its water extract.

Extraction with Acid. As in the ordinary nitroglycerin dynamites, the substances removed from "permissible" explosives by acid extraction are chiefly substances added as antacids, including calcium carbonate, magnesium carbonate and zinc oxide. Other acid-soluble materials that may be present include metallic aluminum or zinc, ferric oxide, manganese dioxide, and calcium silicide. When starch is present, the residue from the water extraction is subjected to hydrolysis in boiling dilute HCl as already described (page 1378), and the acid-soluble inorganic components determined in the filtrate by the usual methods. An extraction with cold acid is made only when there is no starch present.

Extraction with Acetone: Nitrocellulose and Nitrostarch. If either nitrocellulose or nitrostarch is present, an extraction with acetone is made as described for gelatin dynamite (page 1401). It should be noted in connection with the preceding steps in the analysis that in order to avoid partial solution of these substances in ether, the ether used in the ether extraction should be alcohol-free (distilled from sodium), and also that all drying of residues containing these materials should preferably be conducted at 70° instead of 100°, in order to avoid partial decomposition. It is impracticable to separate nitrostarch from nitrocellulose but they are not likely to be found together in the same explosive. Small amounts of nitrocellulose are detected less readily than nitrostarch, which is easily identified by the microscope.

Insoluble Residue and Ash. The insoluble residue is usually carbonaceous combustible or absorbent material and is in most cases readily identified by means of a microscope (preferably binocular) with low power (25–50 diameters). The possible presence of any inorganic material which may have been over-looked in the analysis is detected by means of a determination of ash, the residue being ignited until all carbon is burned off and the mineral residue weighed. This is usually not over 0.2%. If higher than 0.5%, there is reason to suspect that some such material as kieselguhr or clay is present, or that the extractions with water or acid were not complete.

NITROSTARCH EXPLOSIVES

General Nature. Nitrostarch explosives have been for a number of years used to a very considerable extent in this country for commercial blasting purposes, chiefly for quarrying. During the war, explosives of this class were adopted by the United States for certain military purposes and proved satisfactory substitutes for trinitrotoluene as bursting charges for hand grenades, rifle grenades and trench mortar shell.

The commercial nitrostarch explosives may contain, in addition to the nitrostarch, any or all of the following components: oxidizing agents, as sodium or ammonium nitrates, combustible material, such as charcoal, flour, sulphur, etc., mineral oil, and antacids, such as calcium carbonate, or zinc oxide. Nitrostarch military explosives may consist of some such mixture as the above, or may be composed almost entirely of nitrostarch with the addition of relatively small amounts of oils and materials used for granulating. Some types of nitrostarch explosives proposed for military use contained water in amounts up to 10–15% combined with a mixture of nitrostarch and a soluble nitrate.

For commercial use, nitrostarch explosives are put up in cartridge form in paper wrappers, just as nitroglycerin explosives are prepared.

Moisture. Moisture is determined as in the case of nitroglycerin dynamites. 3 to 5 grams of the sample is desiccated over sulphuric acid for 2–3 days (2 days is usually sufficient to give constant weight), or for 24 hours in a vacuum desiccator with a vacuum of at least 700 mm. of mercury. Unlike nitroglycerin, nitrostarch does not volatilize and may be desiccated to constant weight. When heated to higher temperatures, *e.g.*, 100° C., for any extended time, it undergoes, like all nitric esters, a gradual decomposition with loss of weight.

Extraction with Petroleum Ether: Oils, Sulphur, etc. The small amount of alcohol usually present in ordinary grades of ethyl ether, is sufficient to cause partial solution of the nitrostarch, if ethyl ether is used for removing oily ingredients, sulphur, etc., nitrostarch being readily soluble in mixtures of ether and alcohol. It is therefore advisable to use petroleum ether, which does not dissolve nitrostarch, for this extraction.

A sample of about 10 grams of the explosive, in a Gooch crucible with asbestos mat, is extracted with pure petroleum ether of about 0.65 specific gravity, the excess solvent removed by suction and the crucible with sample dried to constant weight at approximately 70° C. The % loss of weight, minus the moisture content, already determined, represents the percentage of ether-soluble material present. The petroleum ether is removed by evaporation, the residue of ether-soluble materials dried, weighed, and its components determined by the methods used for dynamites.

Extraction with Water: Nitrates, Gums, etc. The dried and weighed residue insoluble in petroleum ether is extracted with distilled water to remove the nitrates or other water-soluble materials. The insoluble residue left in the crucible is dried at a temperature of 80° C. (100° may cause some decomposition of the nitrostarch) for several hours, and weighed, the loss of weight being total water extract and serving as a check against the sum of the components separately determined.

Ammonium nitrate, if present, is determined in the water solution by the usual method of distillation of the NH_3 after adding an excess of alkali. Sodium nitrate is determined by evaporating the water extract, volatilizing ammonium salts, and weighing the residue after re-oxidizing with nitric acid if charring has indicated the presence of any water-soluble organic material (page 1378).

If the original explosive is granular in form, the presence of a binding or agglutinating material in the water extract may be suspected. Although numerous other substances may be used for the purpose, gum arabic is frequently employed as a binding agent in different types of dry explosives. A qualitative test for gum arabic has been mentioned on page 1385, and its quantitative determination may be conducted as described on page 1388, by precipitation with basic lead acetate solution.

Insoluble Residue: Nitrostarch, Charcoal, Cereal Products, etc. *Starch.* A microscopic examination of the weighed insoluble residue will usually serve to identify its components. Any un-nitrated starch or cereal products is readily distinguished from nitrostarch by treating with a drop of KI solution of iodine and examining under the microscope, when the un-nitrated starch granules will appear blue or black and the nitrated starch colorless or yellow. Charcoal is identified by its color.

Un-nitrated starch, if present in an amount greater than a trace, is determined by boiling with dilute H_2SO_4 or HCl until iodine solution no longer colors a drop of the liquid blue, then filtering and washing thoroughly. The residue is dried at 100° and weighed, the loss of weight representing starch.

Nitrostarch. Another portion of the insoluble residue (the analysis being conducted in duplicate), is extracted with acetone in a Wiley extractor or other continuous extraction apparatus, or by transferring the residue from the crucible to a small beaker, digesting in acetone with stirring, and filtering through the same crucible, washing with fresh acetone. This extraction dissolves all of the nitrostarch, leaving any charcoal or cereal products that may be present. The residue is dried at 100° and weighed, the loss of weight representing nitrostarch.

The nitrogen content of the nitrostarch may be determined, if desired, by precipitating the nitrostarch from the clear acetone solution by the addition of water and evaporation on a steam bath. A portion of the white, floury precipitate is then dried at 70°–80° C., weighed, and its nitrogen content determined in the nitrometer (see page 354 Vol. I).

Charcoal. If charcoal is present its weight may be taken direct, in the absence of cereal products or other substances. When the residue contains cereal products, the material left after hydrolysis of the starch and extraction of the nitrostarch will contain the crude fibre of the cereal together with charcoal or other insoluble ingredients. A separation of such components is usually impracticable.

Trinitrotoluene (TNT.)

Trinitrotoluene, commonly designated in this country by the abbreviation TNT., is also known in this and other countries by such names as triton, trotyl, tolite, trilite, trinol, tritolo, etc. The term trinitrotoluol, which is probably more commonly used than trinitrotoluene, is incorrect according to approved chemical nomenclature.

This explosive is of the greatest importance as a high explosive for military use, being adaptable as a bursting charge for high explosive shell, trench mortar shell, drop-bombs, grenades, etc., because of its powerful explosive properties, relative safety in manufacture, handling, etc., its stability, its lack of hygroscopicity, and absence of any tendency to form sensitive compounds with metals.

It is classified by the Ordnance Department, U. S. A., into three grades, according to purity—Grades I, II, and III, with solidification points of at least 80.0°, 79.5°, and 76° C., respectively. Other requirements—the same for all grades—are as follows: ash, not more than 0.1%; moisture, not more than 0.1%; insoluble, not more than 0.15%; acidity, not more than 0.01%.

Solidification Point. The determination of the solidification point or "setting point" of TNT. is the best single test for purity of this compound, and is preferably carried out as follows:

A sample of about 50 grams of TNT. is placed in a 1"×6" test tube and melted by placing the tube in an oven at about 90° C. The tube is then inserted through a large cork stopper into a larger test tube about $1\frac{1}{2}''\times7''$, which, in turn, is lowered into a wide-mouth liter bottle, so that the rim of the large tube rests on the neck of the bottle. The inner test tube is provided with a cork stopper containing 3 openings—one for a standard thermometer graduated in 1/10° C., one for a short thermometer which is passed just through the stopper and is used for noting the average temperature of the exposed mercury column of the standard thermometer, and the third opening being a small v-shaped notch at the side of the stopper, through which passes a wire whose lower end is bent in a loop at right angles to the axis of the tube and which is used for stirring the molten sample of TNT.

The standard thermometer is so adjusted that its bulb is in the center of the molten mass, and the stirrer is operated vigorously, the thermometer being watched carefully as the temperature falls. The temperature will finally remain constant for an appreciable time and then rise slightly, owing to the heat of crystallization of the TNT. As this point is reached, readings should be taken about every 15 seconds until the maximum temperature of the rise is reached. This temperature will usually remain constant for several minutes while crystallization is proceeding. The maximum reading, corrected for the emergent stem of the thermometer, is taken as the solidification point of the sample.

Ash. About 5 grams of TNT. is moistened with sulphuric acid and burned in a tared crucible. The residue is again moistened with a few drops of nitric acid and sulphuric acid and again ignited and the resulting ash weighed.

Moisture. A sample of about 5 grams spread on a watch glass is desiccated over sulphuric acid to constant weight.

Insoluble. A sample of about 10 grams is treated with 150 cc. of 95% alcohol, heated to boiling, and filtered while hot through a tared Gooch crucible with asbestos mat. The insoluble residue is washed with hot alcohol, dried at 100° C. and weighed.

Acidity. A 10-gram sample is melted in a large test tube or a flask and shaken with 100 cc. of neutralized boiling water, cooled and the water decanted. A similar treatment is given using 50 cc. of boiling water, the two portions of water combined, cooled and titrated with tenth normal NaOH, using phenolphthalein indicator. The acidity is calculated as % H_2SO_4 in the original sample.

Nitrogen. Nitrogen is not usually determined in the inspection of TNT. but when necessary it may be determined by the Dumas combustion method or the modification of the Kjeldahl method described on page 1382.

Picric Acid

Ordnance Department, U. S. A., specifications for picric acid prescribe that it shall have a solidification point of not less than 120° C.; that it shall contain not more than the following amounts of impurities:

Moisture—0.2% for dry material..................12.0% for wet.
Sulphuric acid (free and combined)............... 0.10%
Ash... 0.2%
Insoluble in water............................... 0.2%
Soluble lead..................................... 0.0004%
Nitric acid (free)............................... none

Solidification Point. Dry the sample at a temperature not exceeding 50° C. Melt sufficient to give a 3-inch column in a 6-inch$\times\frac{3}{4}$-inch test tube immersed in a bath of glycerin heated to 130° C. When the sample is completely melted remove the tube from the bath and stir the sample with a standardized thermometer graduated in 0.10 degrees, until the picric acid solidifies. During solidification the temperature will remain constant for a short time and then undergo a slight rise. The highest temperature reached on this rise is recorded as the solidification point. The test may be more accurately carried out using the apparatus and method as described under trinitrotoluene (p. 1391).

Moisture. A weighed sample of about 10 grams is spread evenly on a tared watch glass and dried to constant weight (about 3–4 hours) at 70° C.

Sulphuric Acid. About 2 grams is weighed and dissolved in 50 cc. of distilled H_2O, acidified with HCl and heated to about boiling. Hot $BaCl_2$ solution is added with stirring and the mixture allowed to stand at least 1 hour on the steam bath. Filter hot on a tared Gooch crucible, wash with water, dry at 100° C. and weigh. Calculate $BaSO_4$ found as H_2SO_4 in original sample.

Ash. About 1 gram is weighed in a platinum crucible, saturated with melted paraffin, burned carefully, and the residue ignited to burn off all carbon. The resulting ash is cooled and weighed.

Insoluble in Water. 10 grams of the sample is treated with 150 cc. boiling water, boiled for 10 minutes, filtered while hot through a tared Gooch crucible, washed well with hot water, and the insoluble residue on the filter dried at 100°, cooled, and weighed.

Soluble Lead. The presence of soluble lead in picric acid is highly objectionable, because lead picrate is an extremely sensitive explosive and its presence would greatly increase the dangers involved in handling and loading picric acid. A weighed sample of about 300 g. is digested in a 2-liter flask with 100 cc. of a hot saturated solution of barium hydroxide in 65% alcohol. 1400 cc. of 95% alcohol is then added and the digestion continued at a temperature below the boiling point (with reflux condenser), until everything except traces of insoluble matter is in solution. The picric acid is then allowed to crystallize on cooling, and the solution filtered off, decanting the clear liquid from the crystals until 500 cc. of filtrate is obtained. This 500 cc., representing 100 g. of picric acid, is treated with 5 drops HNO_3 and 10 cc. of 1% $HgCl_2$ solution, and H_2S passed through it for 15 minutes. Allow the precipitate to settle for 20 minutes, filter and wash with alcohol saturated with H_2S. Dry and ignite the precipitate, then dissolve the residue in 9 cc. of HNO_3 (sp.gr. 1.42) by warming, add warm water to bring the volume to 50 cc., and electrolize at 0.4 ampere and 2.5 volts, temperature 65° C., for 1 hour. Wash the electrode by replacing the beaker with another one containing distilled water without interrupting the current. Dry and weigh the previously tared anode. The weight of lead peroxide found×0.8661 gives the percentage of soluble lead found.

Nitric Acid. No coloration should result when a water solution of picric acid is treated with a solution of diphenylamine in sulphuric acid.

Ammonium Picrate

Ammonium picrate, also known in this country as "Explosive D," is of importance as a military explosive more on account of its insensitiveness to shock and friction, than because of its explosive strength, which is less than that of TNT. Its chief use is as a bursting charge in armor-piercing projectiles.

Military specifications require it to be prepared from picric acid of standard purity, to contain not less than 5.60% ammoniacal nitrogen, and not more than the following amounts of impurities:

Moisture	0.20%
Sulphuric acid (free and combined)	0.10%
Nitrates	trace
Insoluble material	0.20%
Ash	0.20%
Nitrophenols	0.50%

Moisture. A sample of about 10 grams spread on a tared watch glass is dried at 95° C. to constant weight (about 2 hours).

Sulphuric Acid. About 5 grams is dissolved in 100 cc. of hot water, filtered, washed with 25 cc. hot water, the filtrate acidified with HCl, heated to boiling and treated with hot $BaCl_2$ solution. Any precipitate is filtered on a weighed Gooch crucible, dried at 100° C., weighed and calculated as H_2SO_4.

Nitrates. A water solution of the sample tested with diphenylamine and H_2SO_4 should give no blue coloration.

Insoluble Material. A 10-gram sample is boiled with 150 cc. of water for 10 minutes, filtered on a Gooch crucible, the residue washed with hot water, dried at 100° and weighed.

Ash. A sample of about 1 gram is saturated with melted paraffin and burned in a tared crucible, the residue ignited to burn off all carbon, and the ash weighed.

Nitrophenols. 10 grams of powdered sample is treated with 50 cc. of chloroform for 30 minutes with frequent stirring and filtered into a 100 cc. tared flask, the residue being washed with 25 cc. of chloroform. The filtrate is evaporated to dryness and any residue obtained weighed. This residue is treated with ammonium hydroxide, again evaporated to dryness and extracted the second time with 25 cc. of chloroform. The chloroform filtrate is evaporated to dryness and the residue weighed. The difference in weight between this residue and the first residue equals the nitrophenols, other chloroform-soluble having been eliminated by the ammonia treatment and second extraction.

Tetryl

Tetryl is the commercial term applied to the explosive trinitrophenyl-methylnitramine, also improperly called tetranitromethylaniline. Its chief use is as a "booster" charge in high explosive shell, where it serves to transmit the detonating wave from the detonator or fuze to the less sensitive bursting charge. Being in immediate contact with the fuze it must be of a high degree of purity, and is required by Ordnance Department specifications to have a melting point of at least 128° C. and to contain not more than the following amounts of impurities:

Moisture	0.05%
Acidity (as H_2SO_4)	0.01%
Insoluble in acetone	0.30%
Ash	0.15%
Sodium salts	trace

Melting Point. The sample to be used for this test is dried overnight in a vacuum desiccator and pulverized to pass a 100-mesh screen. A capillary melting-point tube is filled to about $\frac{1}{4}$ inch from the bottom and attached to the stem of a standard thermometer so that the sample is next to the center of the bulb. The bath is properly agitated and provision made for correcting for the emergent stem of the thermometer. The temperature of the bath is raised rapidly to 120° C., then at the rate of 1° in 3 minutes, the temperature at which the first meniscus appears across the capillary tube being noted as the melting point.

Moisture. A sample of about 10 grams is weighed in a wide shallow weighing bottle and dried over sulphuric acid in a desiccator for 24 hours, the sample being spread uniformly so that its depth is not over 0.5 cm. The loss of weight is regarded as moisture.

Acidity. A 10-gram sample, finely powdered, is shaken for 5 minutes with 50 cc. of boiled distilled water, filtered, washed with 50 cc. more water, and the filtrate and washings titrated with N/50 NaOH solution using phenolphthalein indicator.

Insoluble in Acetone. 10 grams of sample is dissolved in 75 cc. of acetone, filtered through a tared Gooch crucible, and the residue washed with 25 cc. of acetone, dried to constant weight at 100° C. and weighed.

Ash. The dried residue insoluble in acetone is ignited, cooled in a desiccator and weighed.

Sodium Salts. Any sodium present in tetryl is combined as sodium picrate. 10 grams of the tetryl are boiled in 50 cc. distilled water, cooled, filtered, the filtrate acidified with acetic acid and evaporated to 10 cc., cooled again and filtered. The filtrate is made alkaline with ammonia and treated with 5 cc. of 10% solution of ammoniacal copper sulphate. Any sodium picrate will be precipitated as crystals of cupro ammonium picrate on standing for a few minutes.

Mercury Fulminate

In commercial blasting caps and electric detonators mercury fulminate is generally found intimately mixed with potassium chlorate. It is, however, used without admixture in certain types of detonators, in the fuzes of high explosive shell and for other military purposes. It is usually purchased under specifications which provide that it shall be at least 98% pure, shall be free from acid, and contain not more than 2% insoluble matter, 1% free mercury, and 0.05% chlorine in the form of chlorides.

Preparation of Sample. Mercury fulminate being packed and handled in a thoroughly wet condition until dried just before use, it is generally necessary to dry the sample before testing. This may be done by exposing in a low temperature oven at not more than 50° C. until practically dry, then in a desiccator (not a vacuum desiccator) over sulphuric acid or calcium chloride until its weight is constant.

Mercury Fulminate Content. Exactly 0.3 g. is weighed into a wide-mouthed Erlenmeyer flask containing 250 cc. distilled water, and 30 cc. of a 20% solution of purest sodium thiosulphate is added quickly and the mixture shaken for exactly 1 minute. At once titrate with N/10 hydrochloric acid using 3 drops of methyl orange indicator, the titration to be commenced 1 minute after adding the sodium thiosulphate, and to occupy not more than 1 minute additional time.

The percentage of mercury fulminate is calculated from the volume of standard acid required, after deducting the volume of acid required for a blank determination. Four molecules of HCl are equivalent to 1 mol. of mercury fulminate, or 1 cc. N/10 HCl equals 0.00711575 g. mercury fulminate. The reaction is assumed to be as follows:

$$HgC_2N_2O_2 + 2Na_2S_2O_3 + 2H_2O = HgS_4O_6 + 4NaOH + C_2N_2.$$

Acidity. A 10-g. sample is extracted with 2 successive 25-cc. portions of boiled distilled water in a Gooch crucible, and 3 drops of methyl orange solution (1 g. per liter) added. No red tinge of color should be obtained.

Insoluble Matter. A 2-g. sample is dissolved in hot 20% $Na_2S_2O_3$ solution, filtered through a tared Gooch crucible and any insoluble washed with water then with alcohol and finally with ether, dried at 60°–70° C. and weighed.

Free Mercury. The residue of insoluble matter obtained as described above is treated with a solution of 3 g. KI and 6 g. $Na_2S_2O_3$ in 50 cc. H_2O by passing the solution through the Gooch crucible. Any organic mercury compounds are thus converted into mercuric iodide, which is soluble in $Na_2S_2O_3$ solution. The metallic mercury remains behind on the filter, and is washed with H_2O, dried 1 hour at 80°–90° C., and weighed.

Chlorides. A 5-g. sample of fulminate is extracted in a Gooch crucible with 2 successive 25 cc. portions of distilled water at 90°–100° C. Three drops of strong HNO_3 and 10 drops of 10% $AgNO_3$ solution are added to the filtrate. If a turbidity results, the AgCl should be determined gravimetrically or a fresh sample extracted and the filtrate titrated with a standard $AgNO_3$ solution.

Blasting Caps and Electric Detonators

Preparation of Sample. In the examination of blasting caps or detonators for either commercial or military use, the removal of the detonating composition from the copper or brass shell requires considerable precaution. Blasting caps are emptied by squeezing the cap gently in a pair of "gas forceps," the jaws of the forceps being passed through a small opening in a piece of heavy leather, rubber belting, or similar material, about 6″ square, which serves as a shield to protect the hand in case of the possible explosion of the cap in squeezing. After each squeeze, the loosened portion of the charge is shaken out on a piece of glazed paper, the cap turned slightly in the forceps and again squeezed. The pressure on the cap should be just sufficient to slightly dent it, and in shaking out the charge, the cap should not be tapped on the table or other surface. With these precautions there is little danger of an explosion.

Electric detonators are opened by first cutting off the wires or "legs" close to the shell, then tearing off the upper portion of the shell by means of pointed side-cutting pliers, the cap being held firmly in the fingers and a thin strip of the copper shell being torn off spirally by nipping the top edge of the shell with the forceps. This must be done with great care, especially as the portion of the shell containing the fulminate charge is approached. When the greater portion of the plug which holds the wires in place has been exposed, the plug and wires are gently pulled out, care being taken to avoid force and possible friction, and any adhering particles of the charge brushed off onto glazed paper. The charge is then removed from the lower part of the shell just as in the case of blasting caps.

The charge is removed separately from several of the caps or detonators and each weighed in order to determine the average weight of charge as well as variation of same.

"Reinforced" caps, or those which contain a small perforated inner copper capsule pressed on top of the charge, must be opened in the manner described for electric detonators, in order to remove the inner capsule. Detonators of this type usually contain a main charge of some nitro compound superimposed by a layer of mercury fulminate, mixture of fulminate and chlorate, or lead azide. Although a clean mechanical separation of the two layers is usually not possible, portions can be taken from each and identified by qualitative tests before proceeding with a quantitative examination.

Moisture. The moisture content of the composition is determined by desiccating to constant weight over sulphuric acid or calcium chloride.

Analysis of Composition Containing Mercury Fulminate and Potassium Chlorate. About 2–3 grams of the well-mixed composition is weighed in a Gooch crucible provided with asbestos mat or disc of filter paper or silk, and first moistened with a few drops of alcohol, then extracted with 200–250 cc. of cold water in 15–20 cc. portions, using slight suction after each portion has remained in the crucible for a few minutes. The residue in the filter is dried to constant weight at 60°–70° C. (2–3 hours), and weighed.

The water extract contains the potassium chlorate and a portion of the mercury fulminate, which is slightly soluble in cold water. It is treated with 2 cc. of ammonium hydroxide and H_2S passed to completely precipitate the dissolved mercury fulminate as HgS. This black precipitate is filtered off, washed, dried and weighed. Its weight$\times 1.22$ gives the amount of mercury fulminate dissolved by the water. This weight added to the weight of the dried residue insoluble in water gives the total weight of mercury fulminate in the sample. The $KClO_3$ is found by subtracting the % of mercury fulminate $+\%$ moisture from 100%.

Analysis of Compositions Containing Nitrocompounds. Trinitrotoluene, tetryl or picric acid can be identified by melting point test, TNT. melting at about 79°–80° C., tetryl at about 128° C., and picric acid at about 120°–122° C. They may be extracted from the mixture by means of ethyl ether, in which mercury fulminate is only very slightly soluble, and the determination of $KClO_3$ and mercury fulminate then made as described in the preceding paragraph.

If the main charge is an organic nitrate such as nitrated vegetable ivory, nitrostarch, etc., such material will be left with the mercury fulminate in the insoluble residue after extraction with water. The mercury fulminate is then extracted by means of a hot 20% solution of sodium thiosulphate, leaving the organic nitrate in the Gooch crucible. These materials in the detonating composition can be readily identified by microscopic examination.

In detonators where TNT. or tetryl compose the main portion of the charge, a small amount of lead azide, with or without mercury fulminate may be used as a priming charge for the purpose of initiating the detonation of the nitrocompound. It should be identified in the top portion of the charge, next to the reinforcing cap, and will in all probability be present if mercury fulminate is not found. It is practically insoluble in water and in ether, and will be left in the insoluble residue. If present, fulminate is destroyed by treating the residue, in a flask, with 25 cc. of KOH solution. This converts the lead azide to potassium azide, KN_3. A slight excess of H_2SO_4 is added and the mixture distilled, the distillate, containing HN_3, being collected in water. Enough NaOH is added to the distillate to give an alkaline reaction with litmus, then a little $Pb(NO_3)_2$, when lead azide, PbN_6 will be regenerated as a white precipitate, which may be filtered off, washed with water, then with alcohol, dried in the air, and tested by striking a small portion with a hammer.

Primers

Variations in Composition. Many varieties of composition are used in primers for small arms ammunition, and for other military purposes. The composition must be ignited by the impact of the firing pin, and must give a flame of sufficient intensity and duration to ensure proper ignition of the propellant or of the detonator, depending on the purpose for which the primer is employed. As primers are used with various kinds and granulation of explosives, a priming composition suitable for one purpose is unsuited for another; hence there are many types of priming compositions, a few of which are indicated in the following table:

TYPES OF PRIMER COMPOSITIONS
APPROXIMATE COMPOSITION (PER CENT)

Ingredients	No. 1	No. 2	No. 3	No. 4	No. 5	No. 6	No. 7	No. 8	No. 9	No. 10
Mercury fulminate	31	25	11	..	..	..	..	..	..	28
Potassium chlorate	38	38	53	60	50	51	53	53	47	14
Sulphur	..	..	..	7	3	9	..	..	22	..
Powdered glass	..	..	..	..	..	12	..	..	..	35
Lead sulphocyanate	..	..	..	..	..	..	25	25	..	..
Copper sulphocyanate	..	..	..	3	..	..	..	..	..	..
Barium nitrate	..	6	..	..	..	..	..	..	..	..
TNT	..	..	..	..	..	..	5	..	..	..
Tetryl	..	..	..	..	..	..	..	3	..	..
Antimony sulphide	31	31	36	30	44	26	17	17	31	21
Lead oxide (PbO)	..	..	..	..	..	..	..	2	..	..
Shellac	..	..	..	..	..	2	..	..	..	2
Black powder (meal)	..	..	..	..	3	..	..	..	..	..

In addition to these ingredients most priming compositions are mixed with small amounts of some binding material dissolved in water or alcohol, such as gum arabic, gum tragacanth, glue, shellac, etc. These traces of binding materials are usually disregarded in the analysis of the compositions.

Preparation of Sample. If the caps contain anvils, these must first be carefully removed, as well as any covering of tin foil or paper. The primer composition is then carefully removed from a number of primers and weighed to determine the average charge. It is then carefully crushed, a little at a time, and the sample well mixed. If necessary, it may be removed from the caps by the aid of water or alcohol and the latter removed by evaporation before weighing.

Qualitative Examination. The following special tests may be made use of in connection with a qualitative analysis of the mixture:

A small amount is burned between two watch glasses, the formation of a mirror indicating mercury, antimony, copper or lead. The mercury mirror is readily volatile on gentle ignition.

Extract a portion of the mixture with ether, then with water, then with $Na_2S_2O_3$ solution, then with aqua regia, retaining each of these solutions.

TNT. or tetryl may be present in the ether solution and are identified by m.p. or color test, TNT. giving a deep red color with acetone and KOH. Sulphur is detected by burning a portion of the ether-soluble material and noting odor of SO_2.

The water extract is tested for $KClO_3$ by adding H_2SO_4, boiling, and noting odor of chlorine. A portion is treated with HCl and $FeCl_3$, a red color indicating thiocyanate. The usual $FeSO_4$ ring test is made for nitrates. A white precipitate with H_2SO_4 indicates Ba or Pb.

The aqua regia solution is diluted and tested with H_2S for antimony, lead, and copper. If the precipitate is not orange-red, lead or copper are indicated. Dissolve in HNO_3, neutralize with NH_4OH; a blue solution indicates copper, while lead is detected by the formation of a white precipitate with H_2SO_4.

Any material insoluble in aqua regia may be powdered glass or other abrasive material.

Quantitative Analysis. The method of analysis will depend entirely upon the ingredients indicated by qualitative tests. In general, a separation is best effected by successive extractions with ether, water, $Na_2S_2O_3$ solution (to remove fulminate), dilute or concentrated HCl, and aqua regia. The small amount of mercury fulminate present in the water extract may be determined by precipitation with H_2S or by adding 10–15 cc. of thiosulphate solution and a few drops of methyl orange and titrating with N/10 HCl or H_2SO_4 (see page 513 of Vol. I). Other materials in the water and acid solutions are determined by the usual analytical methods.

Nitrocellulose

General. The term nitrocellulose, or more correctly cellulose nitrate, applies to any nitration product of cellulose, ranging from products containing in the neighborhood of 10–11% N, which are used in the preparation of lacquers and other commercial products, to military guncotton with over 13% N. All of these products are undoubtedly mixtures of the various nitrates of cellulose, as indicated by the fact that there is always some material with low nitrogen content, soluble in ether-alcohol, in high nitrogen guncotton, and some insoluble material in the lower nitrated commercial products. It can usually be shown without great difficulty that any nitrated cotton is a mixture of various nitrates of cellulose.

The products of military importance are the insoluble guncotton of high N-content, and the so-called "pyro" or pyrocellulose, soluble in ether-alcohol and of about 12.60% N-content. In testing these products, the characteristics of most importance are content of nitrogen, solubility in ether-alcohol, and stability. Other determinations generally made are solubility in acetone and ash.

Preparation of Sample. If the sample contains a large excess of water, it is enclosed in a clean cloth and the excess water removed by means of a press or wringer. The pressed sample is then rubbed up in the cloth (not with the bare hand) until lumps are removed, then spread on clean paper trays in an air bath at about 35°–40° C. until "air-dry."

Samples for stability tests and nitrogen determination are treated as noted below, the air-dry sample being suitable for determining solubility and ash.

Nitrogen. About 1 to 1.05 g. of the air-dry sample is roughly weighed in a tared weighing bottle, dried at 95°–100° C. for $1\frac{1}{2}$ hours, cooled in a desiccator and accurately weighed. It is then transferred to the generating bulb of a nitrometer (Du Pont modification; see p. 354) using a total of 20 cc. of 95–96% c.p. H_2SO_4. The sample must be dissolved in the acid either in the weighing bottle or in the cup of the generator, before it is drawn into the generating bulb, and both the weighing bottle and the cup of the generator must be thoroughly washed out with the 20 cc. of H_2SO_4, so that none of the sample is lost. The determination in the nitrometer is completed in the usual manner (p. 354), the result being expressed as % N in the dried sample of nitrocellulose.

Solubility in Ether-Alcohol. (a) *Guncotton:* The amount of ether-alcohol soluble material in guncotton being usually not more than 10–12%, the determination may be made by evaporating a clear solution. Two grams of air-dry sample is placed in a clean dry cork-stoppered 250 cc. cylinder, 67 cc. of 95% ethyl alcohol added to thoroughly wet the guncotton, then 133 cc. of ethyl ether (U.S.P. grade, 96%), added and the mixture well shaken. If the mixture of 2 parts ether and 1 part alcohol be added at once to the sample, a gummy mass may result which dissolves with great difficulty, especially if the solubility is unusually high.

The cylinder is now allowed to stand at a constant temperature of usually 20° C. (15.5° C. is sometimes specified). The solubility of nitrocellulose *increases* as the temperature is *decreased*, hence a constant temperature of digestion is important. During the digestion, which requires at least 1 hour, the cylinder must be thoroughly shaken at 5-minute intervals. The cylinder is now allowed to stand for at least 4 hours, until the insoluble portion of the sample has completely settled and the supernatant liquid is perfectly clear.

50 cc. of the clear solution is now drawn off with a pipette, care being taken not to disturb the settled pulp, and evaporated in a weighed evaporating dish on a steam bath, avoiding loss from violent boiling of the ether. When 25–30% of the solution has been evaporated, 10 cc. of distilled water is added slowly and the evaporation continued to dryness. The effect of the water is to leave the residue in a white, brittle or powdery condition, rather than a tough film which would lose its solvent with difficulty.

The dish is finally placed in an oven at 95–100° C. for $\frac{1}{2}$ hour, cooled in a desiccator, and weighed. The weight of the residue, corrected for the residue in the 50 cc. of ether-alcohol and 10 cc. H_2O used, represents the soluble nitrocellulose in 0.5 g. of the guncotton.

(b) *Pyrocellulose:* The solubility of pyrocellulose may be determined in the manner described for guncotton, but owing to the much larger amount of soluble material present, the evaporation of the residue to constant weight without decomposition involves considerable difficulty. Sufficient water must be added to precipitate the soluble nitrocellulose from solution in a stringy or fibrous condition.

The determination is usually conducted by either the volumetric method or the filtration method.

In the volumetric method, one gram of the air-dry sample is covered with 100 cc. of 95% ethyl alcohol and allowed to stand at least 15 minutes with frequent stirring, 200 cc. of ethyl ether is then added with stirring and the agitation continued until solution is complete. The solution is now allowed to stand at least 4 hours with frequent stirring, during at least 1 hour of which time it is to be kept at a temperature of 15.5° C. It is then transferred to a "solubility tube" and allowed to stand for at least 16 hours, in order that the insoluble material may settle completely. The solubility tubes are glass tubes about 30.6 inches long ✕ 1.3 inches inside diameter, tapering at a point 6 inches from the bottom to a constricted portion about 3 inches long and about .375 inch inside diameter. This narrow bottom portion is graduated to read directly the percentage of insoluble material, the value of the graduations having been first ascertained by comparison with results obtained by the filtration method described below. The tubes are made of heavy glass and provided with vented ground glass stoppers. They hold 300 cc. when filled to about 8 inches below the top.

In the filtration method, the solution is prepared and settled in a solubility tube as described above, and the clear liquid removed as completely as possible by means of a narrow siphon tube of glass. Fresh alcohol and ether are then added as before, the tube shaken and allowed to stand again for 16 hours, when the process may be repeated several times, depending on the amount of insoluble material present. After the last decantation, the residue is washed from the tube to a beaker, using as small a quantity of ether-alcohol as possible, and the mixture filtered through a filtering tube consisting of a 1″✕6″ test tube with its lower end drawn out to a taper terminating in a hole about $\frac{1}{8}$″ diameter. In the lower end of this tube is a small plug of previously ignited asbestos. The filtration is facilitated if the greater part of the asbestos is mixed with the insoluble matter and solvent in the beaker, the mixture well stirred and quickly poured into the filtering tube on top of a small plug of asbestos. In this manner, the insoluble matter becomes mixed with the asbestos and the formation of a gelatinous, impenetrable mat in the tube is avoided. After filtering, the tube is washed with fresh ether-alcohol, dried at 40°–45° C. and finally for 1 hour at 100° C., then cooled in a desiccator and weighed. All combustible matter is then removed by careful ignition, and the tube again weighed, the loss of weight being the total insoluble material in the 1-gram sample.

Solubility in Acetone. A 1-gram sample of air-dry pyrocellulose is treated with about 200 cc. of acetone with frequent stirring until all gelatinous matter has dissolved. The solution is transferred to a solubility tube (described above), the volume made up to about 300 cc. with fresh acetone, well shaken, and allowed to settle for at least 16 hours. The graduations on the tube having been checked by gravimetric determinations, the percentage of residue insoluble in acetone may be read direct, or the filtration method described above may be applied.

Ash. One gram of air-dry sample is weighed in a tared crucible, moistened with 10–15 drops of concentrated nitric acid, and digested for 2–3 hours on a steam bath until converted to a gummy mass. The crucible is then heated carefully over a Bunsen burner until the mass is completely charred, then at a red heat until its weight is constant. The residue is the ash of the sample.

Stability Test: Heat test with Potassium Iodide Starch Paper. The "heat test" or KI test, as it is commonly designated, is the test most commonly employed for determining the stability or degree of purification of nitrocellulose, whether guncotton or pyrocellulose. This test, also referred to as the Abel test, depends on the action of oxides of nitrogen liberated by the nitrocellulose under the influence of heat, the gases in contact with the KI-starch paper liberating iodine which colors the starch.

The sample is dried with great care to avoid contamination, in a clean paper tray, at 35° to 43° C., until its moisture is reduced to the amount which will give the minimum heat test, usually 1.5 to 2%. The proper amount of moisture is determined as follows: During the progress of the drying, the sample on the tray is "rubbed up" from time to time, using a piece of clean tissue paper spread over the back of the hand. When the sample begins to adhere to the paper, due to static electricity, a sample of 1.3 g. is weighed into a standard test tube. These tubes are $5\frac{1}{2}$ inches long, not less than $\frac{1}{2}$ inch inside diameter and not more than $\frac{5}{8}''$ outside diameter, made of glass about 3/64 inch (1.2 mm.) thick. As soon as the first sample is weighed, the tray is replaced in the drying oven for 2–5 minutes, a second sample weighed, and this process repeated until a series of 5 samples have been taken, the last sample being completely dry. This series of samples, if properly taken, will cover the range of moisture content giving the minimum heat test. If the sample in the tray appears to have become too dry during the time the weighings are being made, it may be placed in a moist atmosphere for not more than 2 hours; the entire time of drying and making the test must not exceed 8 hours.

The tubes containing the samples are fitted with clean, fresh cork stoppers through which pass a piece of glass rod into the end of which is fused a small piece of platinum wire bent into a hook. The wire is heated in a flame to clean it, a piece of the standard KI starch test paper, $1'' \times \frac{3}{8}''$, attached, taking care that neither wire nor paper are touched with the fingers, and the paper moistened on its upper portion by touching it with a glass rod dipped in a solution of equal volumes of pure glycerin and water. The stoppers are then inserted in the tubes and the tubes placed in a constant temperature water bath, so that they are immersed to a depth of 2.25 inches. The time of placing in the bath and the time of the appearance of the first faint yellowish discoloration of the test paper are noted. The minimum test given by the 5 samples is taken as the result of the test. The discoloration appears at the lower edge of the moist portion of the paper. The temperature of the heat test bath is 65.5° C. (150° F.) for pyrocellulose, and usually 76.5° C. (170° F.) for guncotton. Pyro is usually required to stand a test of 35 minutes, and guncotton 10 minutes.

A standard test paper is absolutely essential, and is prepared as follows:[1]

The paper used in preparing the test paper is Schleicher and Schüll's filter paper 597. This is cut in strips about 6 by 24 inches, and after being washed by immersing each strip is distilled water for a short time is hung up to dry overnight. The cords on which the paper is hung are clean and the room is free from fumes. The washed and dried paper is dipped in a solution prepared as follows:

The best quality of potassium iodide obtainable is recrystallized three times from hot absolute alcohol, dried, and 1 gram dissolved in 8 ounces of distilled water. Cornstarch is well washed by decantation with distilled water, dried at a low temperature, 3 grams rubbed into a paste with a little cold water, and poured into 8 ounces of boiling water in a flask. After being boiled gently for 10 minutes, the starch solution is cooled and mixed with the potassium iodide solution in a glass trough.

[1] Storm, C. G., Proc. 7th Inter. Congress Appl. Chem., 1909; J. Ind. & Eng. Chem., vol. 1, 1909, page 802.

Each strip of filter paper is immersed in the above-mentioned mixture for about 10 seconds and is then hung over a clean cord to dry. The dipping is done in a dim light and the paper left overnight to dry in a perfectly dark room. Every precaution is taken to insure freedom from contamination in preparing the materials and from laboratory fumes that might cause decomposition. When dry the paper is cut into pieces about $\frac{3}{8}$ by 1 inch and is preserved in the dark in tight glass-stoppered bottles, the edges of the large strips being first trimmed off about one fourth inch to remove portions that are sometimes slightly discolored. When properly prepared the finished paper is perfectly white, any discoloration indicating decomposition due to contamination.

Stability Test at 135° C. In addition to the KI starch test, pyrocellulose is usually required to stand a test at 135° C., made as follows:

The sample is completely dried at 42° C., and 2.5 grams placed in each of 2 heavy glass tubes, 290 mm. long, 18 mm. outside diameter and 15 mm. inside diameter, closed with a cork stopper through which passes a hole 4 mm. in diameter. A strip of litmus paper or standard normal methyl violet paper, 70 mm. long and 20 mm. wide is placed in each tube, its lower edge 25 mm. above the sample, which is pressed down to occupy a depth of 2 inches, the walls of the tube being wiped clean with a roll of paper. The tubes are then heated in a constant temperature bath at 134° to 135° C., all but about 6–7 mm. of the tube being immersed in the bath. They are partially withdrawn for examination of the test papers every 5 minutes after the first 20 minutes of heating, and replaced at once. The time required for reddening of the litmus paper or for turning the methyl violet paper to a salmon pink color is noted as the time of the test. A minimum test of 30 minutes is required with the methyl violet paper, and heating is then continued for a total of 5 hours, during which time there should be no explosion.

The standard normal methyl violet paper is prepared as follows:

Preparation of Methyl Violet Test Paper. A solution is prepared containing the following ingredients: pure rosaniline acetate prepared from 0.2500 g. basic rosaniline, .1680 g. methyl violet (crystal violet), 4 cc. c.p. glycerin, 30 cc. water, and sufficient pure 95% ethyl alcohol to make up to 100 cc. This solution is placed in the angle of an inclined deep rectangular glass tray, and large sheets of Schleicher & Schüll filter paper (No. 597) cut in four strips are dipped in it. In dipping, the strip is held by one end and dipped to within $\frac{1}{4}''$ of this end, withdrawing it slowly up the side of the tray so as to remove surplus solution. The strip is then held horizontally and waved to and fro so as to prevent the solution from running and collecting in spots. As soon as the alcohol has evaporated the strip is suspended vertically to dry, and when dry is cut in strips 20×70 mm. These strips are bottled and kept for use in the 135° test.

SMOKELESS POWDER
Nitrocellulose Powders

At the present time the smokeless powder used by all nations is composed of either colloided nitrocellulose alone or a mixture of colloided nitrocellulose and nitroglycerin. All cannon powder used in this country is of the nitrocellulose type, small-arms powders being of both types. The form and size of the grains are of great variety, depending on the arm in which the propellant is to be employed.

Physical tests made in connection with the examination of smokeless powder include the compression test, determinations of average measurements of the grains, specific gravity, gravimetric density, number of grains per pound, and calculation of burning surface per pound.

Chemical tests include determinations of moisture and volatile solvent, diphenylamine used as stabilizer, ash, material insoluble in ether-alcohol and in acetone, and sometimes nitrogen content.

Stability tests include the 135° C. test, the 115° C. test, and the "Surveillance test."

Moisture and Volatiles. A sample of the powder weighing approximately 1 gram, in the form of thin shavings cut from at least 10 grains, or of whole grains if the powder is too small to cut conveniently, is placed in a clean, dried and weighed 250 cc. beaker, 50 cc. of redistilled 95% (by volume) alcohol, and 100 cc. redistilled ethyl ether added and the beaker allowed to stand under a cover-jar with occasional stirring, until the powder is completely dissolved. This usually requires from 1 to 2 days. When all gelatinous particles of the powder have dissolved, the beaker is heated on the steam bath to evaporate a part of the ether, before precipitation of the nitrocellulose with water. The amount of ether to be evaporated is important, since it largely determines the character of the nitrocellulose precipitate. The presence of too much ether causes a fine sandy precipitate; too little causes a gummy, gelatinous precipitate. A fine, flaky, or fibrous precipitate is desirable. The proper amount of evaporation can be best determined by practice; usually the solution may be evaporated to about $\frac{2}{3}$ its original volume before precipitating. When the proper volume is obtained, 50 cc. of water is added from a graduate, with continual stirring, in 5 cc. portions. If a thick gummy precipitate forms, add a little ether until it becomes flaky; then add the remainder of the 50 cc. of water. The heating is continued with stirring, until most of the ether has evaporated, and the beaker is then left on the bath until the precipitate is just dry. It is then placed in the 100° C. oven for 1 hour, cooled in a desiccator, and weighed as rapidly as possible. To facilitate weighing the weights should be placed on the balance pan before the beaker is removed from the desiccator, so that the exact weight can be adjusted quickly. If more than 10 seconds are consumed in this weighing, the error caused by absorption of moisture from the air is an appreciable one. In any event a check weighing should be made after an additional 30 minutes drying at 100° C.

The final weight of nitrocellulose precipitate subtracted from the weight of the original sample represents the weight of moisture and volatile solvent, and is calculated as per cent of the original sample. If the powder contains diphenylamine, this result is corrected by subtracting from it one fourth of the total diphenylamine content, it having been ascertained by actual trial that approximately this proportion of the diphenylamine is volatilized during the evaporation.

Moisture. An approximation to the actual moisture content of the powder can be obtained by drying a sample of not less than 5 whole grains and not less than 20 grams for 6 hours at 100° C., cooling in a desiccator and weighing, the loss of weight being regarded as equal to the hygroscopic moisture in the powder.

Diphenylamine. The content of diphenylamine used as a stabilizer in smokeless powder is most conveniently and rapidly determined by the "nitration method" as follows:

5 grams of the powder in small grains or slices is treated with 30 cc. of concentrated HNO_3 in a 250 cc. beaker, covered with a watch glass and heated on the steam bath until the powder has been completely decomposed. The solution is then cooled and added to 100 cc. of cold distilled H_2O in a second beaker, stirring vigorously, the first beaker being washed out completely into the second, using additional water. This mixture is now heated on the steam bath until the flocculent precipitate has settled and the liquid has a clear yellow color. It is then cooled, filtered through a weighed Gooch crucible, the precipitate dried at 100° C. and weighed. The weighed precipitate is now dissolved by extracting with acetone, the crucible dried and weighed again, the loss of weight being the nitrodiphenylamine produced by action of the HNO_3 on the diphenylamine. This nitrodiphenylamine is a mixture of nitroproducts, and the empirical factor 0.40576 has been determined for converting it to its equivalent in diphenylamine.

Ash. The ash is determined in the manner described for nitrocellulose (p. 1401), the sample being in the form of slices or small grains, and the digestion with HNO_3 continued until decomposition is practically complete, before heating over a flame.

Solubility in Ether-alcohol. One gram of the sample in slices or small grains is dissolved in 150 cc. of ether-alcohol (2 : 1) in the same manner as for the determination of moisture and volatiles, and transferred to a standard solubility tube (p. 1401), washing it in completely with fresh ether-alcohol so as to bring the total volume to 300 cc. The insoluble material is determined as in pyrocellulose (p. 1401).

Solubility in Acetone. This determination is made in the same manner as the solubility in ether-alcohol, described above.

Stability Test at 135° C. This test is made on duplicate samples in the same manner as described for pyrocellulose (p. 1403). The samples weigh 2.5 grams and are in as nearly whole grains as is consistent with this weight of sample, large grains being turned down on a lathe to fit the standard tubes. The samples are required to stand heating at 134°–135° C. for 5 hours without explosion and must not turn the normal methyl violet paper to salmon pink color in less than one hour.

Stability Test at 115° C. This test is also known as the Ordnance Department 115° test, or the Sy test. Five samples each consisting of not more than 10 grams and not less than 2 whole grains of the powder are weighed on watch glasses and heated at a temperature of 115° ± 0.5° C. for 8 hours daily for 6 days, the oven being brought each day to the proper temperature before the samples are inserted, the samples being allowed to stand at room conditions overnight. At the end of the sixth day's heating, the samples are cooled in a desiccator and weighed. The total loss of weight is regarded as an index of the stability, and must not exceed a specified limit for each particular size of grain.

"Surveillance Test" at 65.5° C. Three samples of approximately 45 grams of powder in whole grains, or, in the case of very large grains, 5 whole grains, are placed in 8-ounce wide-mouth glass stoppered bottles, the stoppers having been previously ground so as to fit tightly. These bottles are then heated in a constant temperature magazine at 65.5° ± 2° C. They are observed several times daily and the time noted when visible fumes of oxides of nitrogen appear in any bottle. The number of days which powder is required to stand this test depends on the web thickness of the grain, and varies from 70 to 140 days. The test is therefore not a laboratory test, but one which more nearly approaches service conditions. It is of great value as an indication of the possible "stability life" of the powder in service.

Nitrogen. The determination of nitrogen in smokeless powder is not usually necessary, in as much as the powder is usually made from nitrocellulose of known nitrogen content, but when desired the determination is made as follows:

An average sample of about 5 grams of the powder in slices or small grains is dissolved in acetone (100 cc. to each 1 g. of sample). When the sample is dissolved, the solution is added drop by drop, preferably from a burette, to 200 cc. of hot water in a beaker, the beaker being immersed in boiling water so as to maintain its contents at about 90° C. During this addition the hot water is continually stirred with a glass rod, so that the precipitated nitrocellulose forms stringy masses which wrap about the rod. Small accumulations of the precipitate are transferred frequently from the rod to another beaker of hot water to prevent the formation of a colloided mass. When 2 g. or more of the precipitate has been collected and the acetone has been volatilized by the hot water, it is removed from the beaker and dried at 35°–40° C. About 1 g. of this dry precipitate is placed in a tared weighing bottle, dried 1 hour at 100° C., weighed, and transferred to the cup of the nitrometer with sulphuric acid. Part of the acid should be added to the precipitate in the weighing bottle before transferring to the nitrometer in order to avoid loss of the dry precipitate in handling. The determination of N is then completed as in the case of nitrocellulose (page 1400). If the powder contains diphenylamine, a correction is necessary for the amount of diphenylamine retained by the precipitated nitrocellulose. This has been found to be an added correction of 0.15% N in the case of powders containing the usual amount of 0.4% diphenylamine. This correction compensates for the nitrogen which becomes combined with the diphenylamine, converting it to nitrodiphenylamines.

Instead of correcting for the effect of the diphenylamine, the latter may be removed from the precipitated nitrocellulose, after air-drying and before final drying at 100° C., by extraction with pure anhydrous ether. Results are quite accurate if the determination is conducted with proper precaution.

Nitroglycerin Smokeless Powders

Powders of this type are composed mainly of nitrocellulose and nitroglycerin and may contain other organic or inorganic substances, such as vaseline, nitro-substitution compounds, substituted ureas or other flame-reducing or surface-hardening agents, diphenylamine, metallic nitrates, carbonates, etc. The nitrocellulose may be either high-nitration guncotton insoluble in ether-alcohol, as in British cordite, or a low-nitration product soluble in nitroglycerin, as in ballistite, or may be a mixture of the two varieties.

The method of analysis usually employed consists of (1) an extraction of the nitroglycerin, nitrosubstitution compounds, vaseline, and other ether-soluble materials by means of anhydrous ether; (2) an extraction of the water-soluble materials; (3) determination of soluble and insoluble nitrocelluloses by separation with ether-alcohol (2 : 1).

The extraction with ether is usually made in a Soxhlet apparatus, using about 20 grams of the powder in slices or small grains, in a paper extraction thimble. About 4 hours is usually required for complete extraction. The ether extract is evaporated to dryness in a tared glass dish under a bell-jar evaporator (page 1376), and the ether-soluble residue weighed. To determine whether it contains other substances than nitroglycerin, it may be poured in small portions at a time into about 20 cc. of strong nitric acid (40° Be) heated on a steam bath. The oxidizing action of the nitric acid destroys the nitroglycerin, and the mixture is then poured into 50–100 cc. of water. Any vaseline or similar substances separate, together with any nitrosubstitution compounds in their original condition or more completely nitrated, diphenylamine in the form of a nitroderivative, etc.

These materials may be separated with more or less completeness by fractional crystallization from ether or other solvent. The exact method to be followed depends on the nature of the materials present.

The residue insoluble in ether is dried and weighed, and then transferred to an Erlenmeyer flask and digested in warm water until any water-soluble materials present have been dissolved. The mixture is filtered, the residue washed with hot water, dried and weighed. The filtrate containing the water-soluble ingredients is examined by the usual analytical methods for inorganic ingredients.

The nitrocellulose insoluble in water is tested for nitrogen content, solubility in ether-alcohol and solubility in acetone, by the methods already described.

Typical Compositions of Commonly Used Explosives

Black Blasting Powder

Sodium nitrate....................73
Charcoal.........................16
Sulphur..........................11 (Bu. of Mines, Bull. No. 80, p. 19.)

Black Military Powder

Potassium nitrate.................75
Charcoal.........................15
Sulphur..........................10

Typical Dynamite formulas—40% grades (Bu. Mines, Bull. No. 80, p. 21).

	Nitro-glycerin	Nitro-Substitution Com.	Ammonium Nitrate	Sodium Nitrate	Nitro-cellu-luse	Wood Pulp	Calcium Carbonate
40% straight Nitroglycerin Dynamite	40	......	......	44	......	15	1
"40%-strength" Ammonia Dynamite	22	......	20	42	......	15	1
"40%-strength" Gelatin Dynamite..	33	......	......	52	1	13	1
"40%-strength" Low-freezing Dynamite..........................	30	10	......	44	......	15	1
"40%-strength" Low-freezing Ammonia Dynamite............	17	4	20	45	......	13	1

Granulated Nitroglycerin Powder ("Judson Powder',

Nitroglycerin.................................... 5　　　　10
Combustible material†...........................35　　　　26
Sodium nitrate............................ ....60　　　　64

Coal Mining Powders. (Permissible Type)

	I	II	III	IV	V	VI
Nitroglycerin...........................	25	15	10	......	......	10
TNT.................................	......	5	......	5	......	......
Ammonium nitrate.......................	......	......	79	90	94	70
Sodium nitrate.........................	34	35	......	......	......	......
Sodium chloride........................	......	......	......	......	......	9
Wood pulp.............................	15	12	10	......	......	10
Flour.................................	25	17	......	5	......	......
Aluminum powder......................	......	......	......	......	3	......
Charcoal..............................	......	......	......	......	3	......
Calcium carbonate.....................	1	1	......	......	......	1
Zinc oxide............................	......	......	1	......	......	......
Magnesium sulphate, cryst...	......	15	......	......	......	......

* Sometimes contains also flour, cornmeal, sulphur, etc.
† Composed of sulphur, coal, and rosin.

NAME INDEX

A

Abel, 111
Albright, 82
Alvisi, 48
Anderson, 107
Arnold, 35

B

Baskerville, 84
Basset, 58
Beck, 58
Beil, 97
Beilby, 35, 71
Benker, 14
Bergmann, 88, 111
Bernthsen, 53
Bertelsmann, 71
Berthelot, 87
Bichel, 87, 100, 102, 110
Billinghurst, 9
Biltz, 13
Birkeland, 21, 23
Blum, 40
Bock, 83
Böcker, 13, 28
Boissière, 78
Borchers, 58
Borland, 97
Bouchand, 109
Boullanger, 5
Brauer, 28
Brank, 105
Braun, 3
Brialles, de, 96
Briner, 56
Broadwell, 58
Brock, 83
Brown, 96
Brownsdon, 111
Brunswig, 87
Bueb, 36, 76, 78, 80, 81
Bunsen, 84
Bütler, 97
Butterfield, 37

C

Calvert, 35
Campari, 84
Campion, 14
Caro, 13, 17, 29, 35, 36, 43, 53, 61, 78
Carpenter, 80
Caspari, 14
Cassel, 56
Chance, 13
Chattaway, 105
Claessen, 91, 96, 103, 106
Clancy, 66
Colson, 13
Conroy, 71, 75, 76
Cooper, 87
Craig, 14
Crookes, 9
Crossley, 3, 17, 23, 61
Crowther, 83
Cundill, 87
Curtis, 103

D

Davies, 56
De Briailles, 96
De Lambilly, 55
De Sourdeval, 78
Donath, 17, 28, 35, 53, 56
Doulton, 19
Duncan, 37
Dupré, 88, 111
Dyes, 13

E

Eissler, 87, 110
Elmore, 84
Erlenmeyer, 72
Erlwein, 71
Escard, 17
Escales, 87, 90, 91, 104, 110
Ewan, 71
Eyde, 21, 23

F

Fairley, 13, 14
Fedotieff, 14
Feld, 14
Flürsheim, 102
Forrest, 71
Foucar, 56
Fowler, 3
Fraenkel, 53, 57, 105
Frank, 13, 29, 35, 36, 43, 61, 78
Frenzel, 17

G

Gelis, 75, 83
Gerlach, 13
Gidden, 82
Girard, 100
Goldschmidt, 14
Gorianoff, 56
Grüneberg, 40
Günzburg, 83
Guttmann, 20, 87, 97, 110, 111

H

Haber, 17, 22, 23, 25, 26, 53, 54, 55
Hahn, 105
Halske, 13
Hargreaves, 111
Hart, 20
Hasenclever, 84
Heine, 82
Heise, 91
Hellriegel, 4
Hempel, 84
Heslop, 76
Hetherington, 83
Hinly, 105
Höchtlen, 82
Hofmann, 71, 82
Hollings, 96
Hooper, 56
Hough, 97
Hurter, 83
Hyronimus, 106

I

Indra, 17, 28, 35, 53, 56

J

Jellinek, 22
Jones, 111
Junk, 111

K

Kaiser, 13, 58
Ketjen, 37
King, 106
Knox, 17, 53
Koenig, 17, 22, 23, 25, 26
Koltunor, 14
Kühlmann, 28

L

Lagrange, 48
Laine, 12
Lambilly, de, 55
Lennox, 14
Le Rossignol, 53
Lesage, 13
Lewell, 105
Lewes, 107
Lidoff, 84
Linder, 80
Lunge, 14, 30, 35, 47, 53, 56, 71, 84, 97
Lyons, 58
Lyte, 105

M

M'Arthur, 71
MacDonald, 94
Marqueritte, 78
Marshall, 48, 110, 111
Massol, 5
Massot, 97
Mehner, 58
Merriman, 110
Messel, 28
Mettegang, 102.
Mettler, 56
Mikolajzak, 91
Miolati, 48
Mond, 36, 43, 78
Muntz, 12

N

Nathan, 87, 88, 89, 90, 92, 95, 96, 111
Naumann, 14
Nernst, 21, 22
Newton, 9
Nithack, 13
Nobel, 87, 97, 98
Norton, 9, 17, 35, 43, 53
Novak, 90, 91
Nydegger, 13, 14

O

Oordt, van, 53
Ost, 71, 78
Ostwald, 13, 28, 29

P

Philips, 110
Pictet, 84
Playfair, 75, 76
Possog, 78

R

Raschen, 75, 83
Readman, 78
Reese, 90
Reid, 97
Resenbeck, 82
Rey, 30
Richards, 53, 57
Rintoul, 88, 89, 90, 91, 108
Robertson, 94, 108, 111
Rossignol, le, 53
Rossiter, 83
Roth, 24, 58
Rowe, 111
Rudeloff, 102

S

Salamon, 82
Sanford, 87, 110
Schachtebeck, 100
Schlutius, 55
Schmidt, 13, 28
Schönherr, 25, 26
Schroetter, 102
Scott, 17, 23, 25, 26, 40, 61
Serpek, 53, 57
Shores, 76
Siemens, 13
Silberrad, 110
Skoglund, 14
Smart, 111
Smith, 82, 84
Södermann, 84
Sourdeval, 78
Spica, 111
Sprengel, 105
Stephenson, 84
Stöhrer, 91
Stollé, 106
Street, 104
Sturzen, 9
Sy, 111

T

Taylor, 37
Tcherniac, 83
Teclu, 104
Tenison-Woods, 14
Thiele, 9, 106
Thilo, 84
Thomson, 20, 87, 88, 89, 90, 91, 92, 96, 97
Traube, 13
Trench, 103
Tschirner, 105
Tucker, 53, 57
Turner, 14

V

Van Oordt, 53
Vincent, 76

W

Wagner, 84
Wahlenberg, 14
Walke, 87
Walker, 37
Watson, 37
Wedekind, 13
Welter, 92
Wendriner, 13
Wesser, 87
Wiernek, 47
Wilfarth, 4
Will, 91, 111
Willcox, 111
Williams, 71, 83
Wilson, 58, 78
Winand, 105
Wislicenus, 106
Witt, 48
Wöhler, 100
Woltereck, 35, 36, 43
Wood, 82

Y

Young, 37, 56

Z

Ziegler, 36

SUBJECT INDEX

A

A2 Monobel, 114
Abel heat test, 92, 111
Absorption towers for nitrogen oxides, 26, 27, 28
Acetone, 107, 108
Acetylene, 75
Acid, nitrating, 88, 93, 101
— waste, 96
After flame rates, 103
Alcohol for mercury fulminate, 105
Alkalsite, 104
Aluminium nitride, 57, 58
— power, 102, 104, 106
Amberite, 114
Amide powder, 88
Amino-acetic acid, 4
Ammonal, 104
Ammoncarbonate, 103
Ammongelatine-dynamite, 98
Ammonia, **35-49, 53-58**
— anhydrous, 49; from aluminium nitride, 57; from cyanamide, 56, 57, 66; from its elements, 53; from nitrides, 57; liquid, 49; nitric acid from, 28, 66; Ostwald's process for, 66; oxidation of, 4, 5, 28, 66; solid, 49; solutions of, 45, 47; sources of, 29, **35-37**; statistics, 36, 37; still for, 39 *et seq.*; synthetic, 28, 37, **53-58**; value of elements in, 55
— dynamite, 112
— nitrate powder, 112
— water, concentrated, 45, 46
Ammoniacal liquors, 38 *et seq.*
Ammonite, 103, 114
Ammonium acetate, 49
— bicarbonate, 48, 55
— bromide, 49
— carbonate, 48
— chlorate, 49
— chloride, 48
— fluoride, 49
— formate, 55
— nitrate, 13, 14, 28, 48, 84; action of heat on, 84; explosives containing, 98, **102-104**; nitrous oxide from, 84
— oleate, 49
— perchlorate, 48, 104
— persulphate, 48
— phosphate, 48
— picrate, 101
— salts, 35-49; fixed, 38; technical, 48, 49; volatile, 38

Ammonium sulphate, 36-48, 55-57; composition of, 38, 41; costs of manufacture, 55, 56, 57; direct processes for, 44, 45; heat of formation, 48; from Mond gas, 43; manufacture, 38-47, 56; price, 38, 55, 56, 57; statistics of, 36, 37
— sulphocyanide, 49, 80, 81
— thiosulphate, 48
Amois, 112
Amvis, 103
Anhydrous ammonia (liquid), 49
Aniline, 110; nitrated, 102
Animal charcoal, 79
Ardeer powder, 98
Artificial silk, 97
Atlas powders, 112
Atmosphere, 3, 21
Azides, 105, 106
Azoimide, 106

B

Bacillus ureæ, 4
Badische Anilin u. Soda Fabrik, 21, 26, 29, 54, 58
Ballistite, 109
Barium nitrate, 13
Bauxite, 57
Bayer & Co., 29
Beaters, for gun cotton, 96
Beetroot sugar waste; see *Schlempe*
Bellite, 103, 113
Benzoic acid, 4
Bergmann, 111
Berlin blue, 81
Bessemer slag, 66
Betaine, 77
Bielby process, 73
Birkeland-Eyde furnace, 23, 24
Bisulphate, sodium, 18, 108
Black powder, 87, 88, 107
Blast furnace gas, 36, 44
Blasting gelatine, 97, 98, 103, 112
— powders, 88
Blue prints, 81
Bobbinite, 87
Bone black, 37
Bones, distillation of, 37
Boron nitride, 58
Brugère's powder, 101, 113

C

Calcium carbide, 61, 63, 78
— chloride, 63, 65

Calcium cyanamide, 29, 56, **61-67**, 78
— fluoride, 63, 65
— nitrate, 12, 13, 27
— nitride, 58
— silicon, 104
Caliche, 9
Cambrite, 114
Cap composition, 111
Carbo-dynamite, 112
Carbon disulphide, 83
Carbonite, 113
Cartridges, dynamite, 100 ; rigid, 103
Cassel Gold Extracting Co., 74
Castner sodamide process, 74
Catalysts, metallic, 54, 55
Caustic ammonia, 45, 46
Celtite, 112
Cementing powder, 67
Ceria, 29, 55
Cerium, 29, 55
Castor oil, use in chlorate explosives, 104
Charge limit, 103
Cheddites, 104
Chemical Engineering Co., 40, 41
Chile saltpetre, 5, 9-11, 88 ; beds of, 9, 10 ; chlorates in, 88 ; use in gunpowder, 88
Chlordinitrobenzene, 102
Chlorate explosives, 104, 105
Chromammonite, 103
Claude plant, 65
Coal, nitrogen in, 35
— gas, 36 ; ferocyanide from, 79 ; purification of, 80, 81, 82 ; prussiate recovery from, 80
Coke, 35, 36
Coke-oven gas, ammonium sulphate from, 44
Colliery steelite, 104
Concentrated ammonia water, 45
Coppée ammonia recovery plants, 41, 42, 45, 46
Copper, for detonators, 105 ; in explosive mixtures, 104
Cordite, 66, 107, 114
Cotton for gun cotton, 92
Cresilite, 102
Crops, rotation of, 4
Cyanamide, 109 ; see also *Calcium Cyanamide*
Cyanate, potassium, 73
Cyanides, 71 *et seq.* ; Erlenmeyer process for, 72 ; from ammonia, carbon and alkali metal, or salt, 73 ; from atmospheric nitrogen, 78, 79 ; from blast furnace gases, 36 ; from coal gas, 81, 82 ; from cyanamide, 66, 78 ; from ferrocyanides, 72, 73 ; from prussiates, 72, 73 ; from schlempe, 76 ; from sodamide, 74, 75 ; statistics, 71 ; from sulphocyanides, 75, 76

D

Dahmenite, 103
Decanitro-cellulose, 96
Denitrification, 5

Depression of freezing point, of nitroglycerine, 91 ; of picric acid, 100
Designollé's torpedo powders, 113
Devil liquor, 41
Dicyandiamide, 66
Diethyldiphenylurea, 109
Diglycerine, 91
Dinitroacetin, 92
Dinitroaniline, 102
Dinitrobenzine, 102
Dinitroformin, 92
Dinitroglycerine, 91
Dinitromesitylene, 102
Dinitromonochlorhydrin, 91
Dinitronaphthalene, 102
Dinitroxylene, 102
Dinitrophenolsulphonic acid, 101
Dinitropseudocumene, 102
Dinitrotoluene, 100, 102, 104
Diphenylamine, 102, 110
Disulphate, sodium, 18
Dodecanitro-cellulose, 96
Dorfite, 103
Doulton's nitric acid condensing plant, 19
Dry ammonia, 49
Dualine, 112
Dynamites, 97, 98, 112

E

E.C. powder, 114
Ecrasite, 100
Electronite, 103
Emmensite, 113
Ennea nitro-cellulose, 96
Erlenmeyer's process for cyanides, 72
Euncanitro-cellulose, 96
Exit gases from ammonium sulphate plant, 41
Explosives, 85-117 ; analysis of, 110, 111 ; composition of, 112 ; permitted, 117 ; statistics, 115-117

F

Faversham powder, 103, 113
Favierite, 113
Feldmann's apparatus, 39
Ferric ferrocyanide, 82
Ferricyanide, potassium, 81
Ferrocyanides, 79-82 ; from coal-gas, 79 ; from sulphocyanide, 75 ; potassium, 79 ; sodium, 79
Ferrodur, 67
Ferrosilicon, 104
Fixation of atmospheric nitrogen, 3, 4, 5, **21-29**, 36, 37, **53-58**, **61-67**
Fixed ammonium salts, 38
Flameless powders, 109
Flesh-eating mammals, elimination of nitrogen from, 4
Forcite, 112
Fulminate of mercury, 105, 111
Fulminic acid, 105
Fume tests, 111

G

Gas water, 38 *et seq.*
Gelatineastralite, 104
Gelatine dynamite, 97, 103, 112
Gelatinewetterastralite, 104
Gelignite, 97, 98, 112
Geloxite, 112
Giant powder, 112
Glycerine polymerised, 91
— dinitrate, 91
Glycocoll, 4
Gold extraction by cyanide, 66, 71, 72
Graphite from cyanamide, 56
Griesheim nitric acid plant, 20
Guanidine, 66
Guhr dynamite, 97
Gun cotton, 92-97, 103
Gunpowders, 87, 88, 103, 107
Guttmann's nitric acid plant, 20
— test, 111

H

Hercules powders, 112
Hexanitrodiphenylamine, 102
Hippuric acid, 4
Hofmann's blue, 83
Hydrogen, industrial, 55

I

Imperial Schultze powder, 114
Iodine, 10
Iron oxide as catalyst, 29

K

Kalkstickstoff, 61, 78
Kieselguhr, 97
— dynamite, 97, 98
King's Norton Metal Co., 106
Kopper ammonia recovery plant, 44
Kynite, 113
Kynoch's smokeless powder, 114

L

Lagrange process, 48
Lead azide, 106
Leather distillation, 37
Leguminosæ, 4
Linde plant, 65
Linseed oil in explosives, 105
Liquid ammonia, 49
Liquor ammonia, 45, 46
Lithium nitride, 58
Lyddite, 100

M

Magnesium, 104
Magnesium nitride, 58
Manganese as catalyst, 55

Melinite, 100, 102
Mercury fulminate, 105, 111
Mesitylene, nitrated, 102
Metanitroaniline, 102
Methylamines, 76
Micrococcus ureæ, 4
Mineral blue, 82
Mines, gun cotton for, 96
Mixing machines, 99, 100
Mond gas, 35, 43, 80
Monobel, 114
Mononitronaphthalene, 100, 104
Mononitrotoluene, 102

N

Naphthalene in explosives, 105
Nitrate, ammonium, 13, 14, 28; see also *Ammonium Nitrate*
Nitrate, calcium, 27; see also *Calcium Nitrate*
— polymerised glycerine, 91
— sodium; see *Sodium Nitrate*
Nitrator separator, 88, 89, 90,
Nitrates, 3, 4, 5, 26
— Products Co., 29
Nitre cake, 18
Nitro acid, 17-31, for guncotton, 93; for nitroglycerine, 88; manufacture from ammonia, 28, 29; manufacture from atmosphere, 21-28; manufacture from Chile saltpetre, 17-20; properties of 29, 30; Ostwald process, 28, 29; red fuming, 30; specific gravity of, 30; statistics, 31; transport of, 30; vacuum process, 20
Nitric organism, 5
— oxide, 21, 22, 26
Nitride, aluminium, 57, 58
— boron, 58
— calcium, 58
— lithium, 58
— magnesium, 58
— silicon, 58
— titanium, 58
Nitrides, ammonia from, 57, 58
Nitrification, 5
Nitrite, sodium, 14, 27
Nitrites, 4, 5, 14, 27
Nitroanilines, 102
Nitrobenzene, 91
Nitrocellulose, 96, 106, 107, 111
Nitrogen, atmospheric, 3; circulation in nature, 3 *et seq.*; in coal, 35; fixation of, 3, 4, 5, 21-29, 36, 37, 53-58, 61-67; manufacture of gaseous, 63, 65
Nitrogen, fixing bacteria, 3, 4, 5
— oxides, absorption plant for, 26, 27
— Products and Carbide Company, 61
— products from cyanamide, 66, 67
Nitrogenous manures, 66
Nitroglycerine, 88-91; depression of freezing point, 91, 92; estimation of, 110; manufacture of, 89-91; nitration of, 88; properties of, 91; smokeless powders containing, 106

Nitroguanidine, 66
Nitrolime, 61, 78
Nitronaphthalene, 100, 104
Nitrophenols, 102
Nitrosococcus, 5
Nitrosomonas, 5
Nitrostarch, 97
Nitrotoluene, 91
Nitrous oxide, 84, 106
Nordhausen sulphuric acid, 93
Norway, water power of, 61
Norwegian saltpetre, 13, 27

O

Octonitrocellulose, 96
Odda, 61
Oil, castor, 104
— linseed, 105
Osmium, catalytic, 54, 55
Ostwald process for nitric acid, 13, 28, 66
Otto-Hilgenstock ammonia recovery process, 45
Oxonite, 113
Oxycellulose, 92

P

Paraffin in explosives, 105
Paris blue, 82
Pauling furnace, 23, 24
Peat, 36, 43
Perchlorate explosives, 104
Percussion caps, 105
Permitted explosives, 117
Permonite, 104
Pertite, 100
Peru saltpetre beds, 9, 10
Petrofracteur, 113
Petroklastite, 88
Phenolsulphonic acid, 101
Phœnix powder, 113
Piano d'orta, 63
Picrates, 101
Picric acid, 100, 103, 106, 111
Plastammone, 104
Plastrotyl, 102
Platinum, catalytic, 28, 29, 55
Poachers for gun cotton, 95, 96
Polyglycerines, 91
Potassium cyanate, 73
— cyanide, 71, 72
— ferricyanide, 81
— ferrocyanide, 72, 73, 79
— nitrate, 11, 12
— sulphocyanide, 76
Powders, smokeless, 97, **106-110**, 114 ; sport-
    ing, 109 ; stability of, 109, 110
Preheaters, 39
Pressure curves of powders, 107
Primers, 105
Priming composition, 101
Producer gas, 36, 44
Prussiates, 71 *et seq.* ; from sulphocyanide,
    75 ; recovery from coal-gas, 80

Prussian blue, 81 ; soluble, 82
Pseudocumene, nitrated, 102
Pulping gun cotton, 95, 96
Pyrites, burnt, use as catalyst, 29
Pyrosulphate, sodium, 18

R

Rack-a-Rock, 113
Rendrock, 112
Rexite, 113
Rhenish dynamite, 98, 112
Rigid cartridges, 103
Roburite, 103, 113
Rotation of crops, 4
Ruthenium, catalytic, 55

S

Safety explosives, 102-104
Sal ammoniac, 48
Saltpetre, Norwegian, 13, 27 ; plantations
    for, 12 ; prismatic, 11
Saturators, 40
Schimose, 100
Schlempe, 36, 72, 76
Schönherr furnace, 25
Schultze smokeless powder, 114
Securite, 103, 113
Serpek process, 57
Sewage, ammonia recovery from, 37
Shale, distillation of, 36
Siepermann's process, 73
Silesia powder, 104
Silicon, 104
— carbide, 104
— nitride, 58
Silk, artificial, 97
Smokeless powder, 97, **106-110**, 114
Société Générale des Nitrates, 58
Sodamide, 74, 106
Sodium bisulphate, 18
— bisulphite, 108
— cyanide, 66, 72 *et seq.*
— disulphate, 17
— ferrocyanide, 79
— industry, 73
— metal, 72
— nitrate, 5, 9, 10, 11, 97
— nitrite, 14, 27
— nitrite nitrate, 28
— pyrosulphate, 18
— sulphide, 18
Solenite, 109
Solid ammonia, 49
Soluble blue, 82
— nitrocellulose, 109
Spermaceti in explosives, 105
Spica's test, 111
Sporting ballistite, 114
— powders, 109
Sprengsalpeter, 88
Stabilisation of gun-cotton, 94
Stabilisers, 110

Stability of powders, 109, 110, 111
— tests, 111
Starch, nitrated, 97
Stills, ammonia, 39, *et seq.*
Stowite, 113
Sugar waste ; see *Schlempe*
Sulphocyanide, ammonium, 80
Sulphocyanides, cyanide from, 75, 76 ; recovery from coal gas, 82 ; synthetic 83, 84
Sulphuric acid for nitrating, 88, 93, 101
Super-rippite, 114
Sweden, water power of, 61
Sy test, 111
Synthetic ammonia, 28, **53-58**
— sulphocyanides, 83, 84

T

Tar in Explosives, 105
Tartrates, 109
Teasing machine, 93
Tetranitroaniline, 102
Tetranitrodiglycerine, 91
Tetranitroethylaniline, 106
Tetranitromethane, 105
Tetranitromethylaniline, 106
Tetranitronaphthalene, 102
Tetranyl, 102
Tetryl, 106
Testing explosives, 103, 111
Thiocyanates, 82
Thorium oxide, catalytic, 13, 29
Titanium nitride, 58
Tolite, 102
Tonite, 113
Torpedoes, 96
Trace tests, 111
Trench's flameless explosive, 103
Trilet, 102
Trinitroaminoanisol, 102
Trinitroaminophenetol, 102
Trinitrobenzene, 102
Trinitrocresol, 100, 101, 102
Trimethylamine, 76, 77
Trinitromesitylene, 102
Trinitronaphthalene, 102
Trinitrophenol, 100
Trinitrophenylmethylnitramine, 106
Trinitropseudocumene, 102
Trinitrotolnene, 96, 100, **101-102**, 103, 106

Trinitroxylene, 102
Trinol, 102
Triplastit, 102
Trotyl, 102
Tungsten, catalytic, 55

U

Ultramarine, 81
Unsafe explosives, 103
Uranium, catalytic, 54, 55
Urea, 4, 5, 66
Urine, ammonia recovery from, 37

V

Valentiner's nitric acid process, 20
Vaseline, 109
Victor powder,
Vielle, 111
Vigorite, 112
Volatile ammonium salts, 38
Vulcan powder, 112

W

Walsrode powder, 114
Waltham Abbey, 88, 92, 94, 95, 108, 111
Waste acids, revivification of 93, 96
Water gas, 35
Westerregeln, 63
Wetter-dynammon, 88
Wetter-fulmenite, 103
Will test, 111
Williamson's blue, 82
— violet, 82
Withnell powder, 113

X

Xylene, 102

Y

Yonckite, 104

Z

Zinc, in explosive mixtures, 104

Printed at THE DARIEN PRESS, *Edinburgh.*

SECOND EDITION, REVISED AND ENLARGED.

Containing over 750 pages and 250 Illustrations.   Royal Octavo
Cloth.   Price 21s. net

Industrial and
Manufacturing Chemistry
ORGANIC

BY

GEOFFREY MARTIN, Ph.D., D.Sc., B.Sc., F.C.S.

Industrial Chemist and Chemical Patent Expert
Member of the Society of Chemical Industry
Formerly Lecturer and Demonstrator in Chemistry at University College, Nottingham

Assisted by Eminent Specialists

::——LIST OF SECTIONS——::

1. **The Oil, Fat, Soap, and Dairy Industries** (by Dr NEWTON FRIEND, Dr GOODWIN, Dr MARTIN, Mr C. A. MITCHELL, Mr W. H. STEPHENS).
2. **The Sugar Industry** (by Dr DRUCE LANDER).
3. **The Starch Industry** (by Dr G. MARTIN).
4. **The Cellulose Industry** (by Mr A. J. CARRIER and Dr G. MARTIN).
5. **The Fermentation Industries** (by Mr C. H. GRIFFITHS, Dr G. MARTIN, Dr A. SLATOR, and Mr W. H. STEPHENS).
   (1) FERMENTS, ENZYMES, BACTERIA. (2) THE RATE OF ALCOHOLIC FERMENTATION BY LIVING YEAST. (3) WINE. (4) BEER. (5) SPIRITS AND INDUSTRIAL ALCOHOL. Parts I. and II. (6) VINEGAR, LACTIC AND BUTYRIC ACIDS.
6. **The Charcoal and Wood Distilling Industry** (by Dr G. MARTIN).
7. **The Turpentine and Rosin Industries** (by Dr G. MARTIN).
   Appendix—NATURAL AND SYNTHETIC CAMPHOR.
8. **Industrial Gums and Resins** (by Dr NEWTON FRIEND).
9. **The Rubber Industry** (by Mr A. J. CARRIER, B.Sc. ; Dr G. MARTIN ; and Mr. G. H. MARTIN).
   (1) NATURAL RUBBER, GUTTAPERCHA, AND BALATA. (2) SYNTHETIC RUBBER.
10. **The Industry of Aliphatic Chemicals** (by Dr G. MARTIN).
11. **The Illuminating Gas Industry** (by Mr E. A. DANCASTER, B.Sc.), including a section on **Acetylene** (by Mr F. B. GATEHOUSE).
12. **The Coal Tar and Coal Tar Product Industry** (by Dr G. MARTIN).
13. **Industry of Synthetic Organic Coloural Matters** (by Mr T. BEACALL, B.A., and Dr G. MARTIN).
14. **Industry of Natural Dye Stuffs** (by Dr G. MARTIN).
15. **The Ink Industry** (by Dr G. MARTIN).
16. **The Pigment and Paint Industry** (by Dr NEWTON FRIEND).
17. **The Textile Fibre, Bleaching, and Waterproofing Industries** (by Dr G. MARTIN).
18. **The Dyeing and Colour-Printing Industries** (by Dr HENRY S. SAND).
19. **The Leather and Tanning Industry** (by Mr DOUGLAS L. LAW, B.Sc., F.I.C.).
20. **The Glue, Gelatine, and Albumen Industry** (by Dr G. MARTIN).
21. **The Industry of Modern Synthetic and other Drugs** (by Dr F. CHALLENGER).
22. **The Modern Explosive Industry** (by Mr WM. BARBOUR, M.A., B.Sc.).
23. **The Industry of Photographic Chemicals** (by Dr G. MARTIN).

LONDON   -   -   CROSBY LOCKWOOD AND SON

www.ingramcontent.com/pod-product-compliance
Lightning Source LLC
Chambersburg PA
CBHW080551220326
41599CB00032B/6438